Khukuri
Kike, History and the Gurkhas

khukuri

Kike, History and the Gurkhas

KAMAN SING KIRANTI RAI
(A Khukuri Guru)
(KKKG)

VAJRA
Publications Inc.Pvt.Ltd.

Published by
Vajra Publications Inc.Pvt.Ltd.
Jyatha, Thamel, P.O. Box 21779, Kathmandu, Nepal
Tel.: 977-1-5320562
e-mail: vajrabooksktm@gmail.com
www.vajrabookshop.com

ISBN: 978-9937-624-23-7

Printed in Nepal by
Dongol Printers
Tel.: 4281682
e-mail: dongolprinters@gmail.com

Never Underestimate small men as:

Napoleon stood only 5 feet 1 inch
Julius Caesar stood 5 feet 2 inches
Alexander the Great stood 5 feet 3 inches

This did not stop them from being military geniuses.

Significantly, KKKG stood 5 feet 3 inches,
and 'here comes the Kike'!

Table of Contents

List of Photos and Drawings

Chapter I

Introduction and acknowledgements

This particular weapon known as kike among Kiranti people, khukuri among Nepali speakers and kukri among foreigners lacks a written history. One finds it very difficult to pin point its exact genesis. Earlier stories relate kike with myths and faiths of the land. Full of myths, legends and bare facts, such stories make one hard to trace down kike's exact origin. Many times, I came to a conclusion that finding its origin is as hard as chaffing the grains from the husks.

I have been studying and suffering much with this task of writing history of kike. I would love to call it kike because the Kirantis, the indigenous community I belong, have called it by this name for ages. Now in my early 80s, it seems as if I am taking a journey of thousands miles by deciding to write a book about kike, a metaphor of the land of warriors. Now a retired British Gurkha, deep down I believe that there is no dish so sweet as that earned by one's own labor, even if it may be the thinnest gruel. I accept this fact that he who is afraid of committing mistake in life stops himself from making discoveries. I request my readers to regard all the mistakes and opinions they may come across in this book as mine alone.

Sometimes, a man travels around the world in search of things he is looking for only to discover that it has been there at his own home for his wonder. I believe that written history such as books and other forms of documents are the testament of advices from dying old persons to rising youths. Or let me put it in my own terms, writing a book is like the orders passed on by the sentry going off duty to the sentry relieving him. This book, I request, should be taken as the same kind of order passed to new generation.

In this book, I point out a geocultural reality about kike or khukuris. Warriors don't come from the sky; they grow from the ambience they are born in. Kirantis have traditions, customs and rites that are directly and indirectly associated with kikes. For us, kike is something in our blood and flesh. As I will be repeatedly writing kike is a distinct weapon used by the Rais and the Limbus as a spiritual object. It is used for several religious acts, and we believe that it possesses certain magical power. We associate it with the soul of Rais and other Kirantis of the eastern part of Nepal.

The Rais and other members of the Kirant community have their own ages-old methods of worshipping as well as handling kike. Kike is honored as 'Veen' or the Lord. This Lord is believed to be residing on stones, mountains, rivers, lakes, jungle, and in and around our homes. Rais also worship the Sun and the Moon. The Sun gives heat and light to all high and low, rich and poor, good and bad irrespective of caste, color and creed. The Moon gives human beings its shining divine pleasure and calmness even during darkness and trouble. This is the major reason why these symbols are marked on Rai's ritually important kike that is safely placed in the prayer room of every Kiranti Rai's home. This sacred kike has a cutting mark in the blade directly in front of kanzo or bolster.

A Kiranti boy usually receives his own kike by his fifth birthday. This is done to inspire him to become a skillful khukuribaaj. He learns to handle it before he is a mature boy. Regarding this very tradition of transforming a Kiranti boy into an enthusiastic khukuribaaj there is an old saying that goes: "Banko kandalai tikharnu pardaina", or wild thorn needs no human help to sharpen it. People in this part of the world believe that Rais and other Kiranti youths become self-trained khukuribaaj. In the hand of a real khukuribaaj or kikebaaj, a kike is equal to any silencer weapons. Kiranti Rais believe that their kike holds the spirit of their warrior ancestors. Kike, therefore, is the lasting memory of ancestor warriors. It is believed to bring pride, charm, good luck and glory to the families.

When a Kiranti leaves his home for a battle, his senior-most family member, or even his dearest wife presents him a religiously, traditionally made kike wrapped in a white patuka or clothe along with a kusume rumal or a white handkerchief. With it, he is offered a garland of flower. They put tika or sacred raw rice on his forehead. For Kirantis, white is the color of power that is believed to generate positive energy, strength and confidence to all the people of the community.

In the ancient time, our ancestors were supposed to have developed a repertoire of knowledge and wisdom associated with livelihood and human existence in this universe, and passed them orally. They called such heritage of knowledge and wisdom the thutur Veda, the form of knowledge that has survived orally. This heritage has been passed through generation to generation as a text of mantras. Such mantras were later called the Mundhum, the holy book of the Kirantis. This holy book contains certain lines or verses, which reveal that our ancestors tried their best to explain the significance of kike to the people. Though potent with meaning, kike for outsiders is simply a weapon.

Explorers, foreigners and traders to Nepal could hardly make a meaning of kike. Back in earlier eighteenth century, Englishmen's ear heard the sacred word khukuri. Thus came into existence words like Kookari or Kookiree, Kukrey, Kukuuri, Koo-ka-ri, Kukri and so forth. Colonel Kirkpatrick said to have been the first English military officer to visit Nepal. In 1793, he praised khukuri in the following manner: the knife worn by every Nepeulian or Nepali called "khookheri"(sic). These days most English dictionaries explain khukuri as a sharp Gurkha knife or a short sword. 'Kike', 'Daabe', 'Kohoch' or Kukri, either way, the thing itself is renowned national weapon of earlier Kirantis and today that of Gurkhas of Nepal. But one should not forget the important reality that kike is a sacred sword of Kiranti people. Once illness strikes, a cure must be found. Beside herbs and medicines, people in Kirant community are advised to satisfy their appropriate deities, ghosts or evil spirits from around. Kirantis use kike in performances of several important rituals and spiritual practices associated with life of their family members.

After a baby is born, his or her umbilical cord is cut either by kike, khurmi or often with a karda. Karda then is placed under the pillow of a newborn's cradle in order to frighten off the evil spirits. Kike and karda are also used for purification, naming, rice feeding, and for baby boy's 'chhewar' or his first hair cutting ceremony.

During a Kiranti's marriage ceremony, the bridegroom is clad in a white daura suruwal an especial Nepali dress, pheta or turban on his head, and importantly his best kike is tucked in his patuka, the cloth belt. Bridegroom places white raw rice of sacred values on the parting of bride's upper forehead. This is called siundo. This ritual marking heralds the completion of marriage that attests the irreversibility of marriage thereafter. Bridegroom's mother uses khurmi or khurumbi to welcome bride from the main door of her home. They perform rituals to assure that from this particular day onwards, they will guarantee bride's care and further protections.

It is ritually warranted that kike is used in funeral procession. As procession starts an unsheathed khukuri is held high in hand by shaman or by any eldest relative of the deceased. Then only, do the group makes its way to burial or cremation ground.

I believe that going for war for a Kiranti is to fall in love with their myths and legends associated with kikes or khukuris. This psychological force has played important role in making the identity of Gurkha warriors across the globe. Khukuri, in fact, is a short-ranged as well as a noiseless weapon. In any form of close-quarter combat, one should be very quick and precise while using it and with a roal. One should not let his foe draw his khukuri for that matter any kind of weapon or else he will be too late. Every Kiranti knows that war is a game of life and death. The teaching of the holy Mundhum goes like this:

Slain, you shall obtain heaven,
Victorious, you will enjoy earth
Therefore, stand up, O sons of Kirantis,
Resolve to fight for right and just!

Rai Spirit of Kike

Because of faith and daily life activities associated with kike, Rais believe that there is something called Rai spirit of kike.

1. Kike carries the pride of the ancient Rai culture and traditions.
2. It represents warrior's spirit of Rai clan.
3. It holds souls of ancestors and families of the Rais.
4. It symbolizes courage and loyalty of the Rais.
5. It honors Gods of the Rai clan and their families.
6. It captures forces of the nature.
7. It embraces power of the universe.
8. It represents spirit of the divine.
9. It purifies warriors' deeds.
10. It appeases spirit of warrior ancestors.
11. It is the temple where their souls abide.

Khukuri Inspires

Khukuri inspires bravery and determination in battle, thus:

a. It infuses courage and bravery in difficult times.
b. It brings quick death to enemy.
c. It frees one from fear, grief and shame.
d. It embraces souls of enemy killed in battle.
e. It is good for right and just attack.
f. Finally, it completes its mission with honor and pride.

Salutation and Kike

a. Bringing kike to holder's forehead means salutation to the spirits of Kiranti warriors of the past, present and future.
b. Bringing kike in front of holder's body at full attention indicates uniting the spirits of all Kiranti warriors.
c. Pointing kike on the ground signifies paying tribute to the fallen Kiranti warriors.

Readers will find that writing a history of khukuri or kike is synonymous with writing the history of Gurkhas. Frankly speaking, I don't want to prove any hypothesis in this book. I simply want to make people know kike or khukuri's power, and the role it has played in maintaining peace and order in this world

I would like to thank the following persons and institutions for inspiring, suggesting and helping me to write this book:

Shree Netra Bahadur Thapa for opening up the mysterious world of kike of the Kirantis in 1990.

Major ND Wylie Carrick MBE 2nd KEOGR/RGR for providing me Dr. M Gyi's address in July 2002. Without this address, I would not have reached to the right person at right time. Meeting rifleman Man Bahadur Rai (Dr. M. Gyi), service number 105598, 3/10 GR was a turning point in my adventure of writing this book. It was him who restored this ancient Kiranti Rai's kike system during the World War II in Burma. He and his ex-kike gurus and staffs had played a major role for providing an honorable status to this weapon. Therefore, I would like to humbly thank Dr. Gyi and his team for their collective effort.

Captain D. F. Harding MA, FRAS, ex- 1/10th GR must be thanked here for providing me some important information about the Gurkha Museum, UK and catalogues of the Khukuris.

I would like to thank captain (Retd.) John Powell from the United States Navy for providing me important information about the London made kukri or khukuri.

Lieutenant Colonel W.J. Dawson OBE RGR's pamphlet on 'Kukri fighting techniques' GTW, UK provided me important information.

Books such as *Nepalese Arms* (An illustrated handbook of Indian arms and of Nepal, Burma, Thailand and Malaya, originally published in London in 1880), *Kukri* the journal of the Brigade of Gurkhas No. 35, May 1983 and the *Soldier* magazine of the British Army have remained as driving force behind this book. This book, my readers will find falls in the academic line created by these books, journals and catalogues.

This book could not have been written without generous help and suggestions provided by many of my great friends. The list is too long to enumerate, but I thank them all with sincerity.

I would like to thank the National Museum, Chhauni, Kathmandu, the Gurkha Museum, England and the National Army Museum, London for kindly giving me permission to quote few passages as well as for obtaining few photographs of the World War I and II.

Thanks to Captain (QGO) Bhairamani Rai (10 GR), WO2 Lalit Kumar Lama (10 GR), Captain (QGO) Mitra Sen Pradhan (QGE) for all the help on providing information, resources, discourses and photos on Khukuri.

Captain DF Harding (MA, FRAS) for providing valuable information and articles on Khukuri and Gurkha and all the correspondence.

I thank my commanding officer Major General R.W.L McAlister (CB, OBE) for the inspiration. He is a dear friend of my uncle, late Major (GCO) Purne Rai.

Bikash Ongchongbo Subba and Gaurav Subba helped me get photos. With them, I discussed the subject of this book. I appreciate their moral support.

Tirtha P. Bhattarai helped me to compile all the materials to develop it into an initial draft. Therefore, I would like to thank him for his help.

I would like to thank the British Council Library and the Kaiser Library at Kathmandu for helping me to get valuable information and resources on kukri, and also providing me valuable and important photographs.

I thank my wife, Sabitri Rai, my dearling daughters Urvashi Rai, Sujata Rai and Barsha Rai, my son Sudarshan Rai for their unconditional love, support and encouragement. They have supported me from the first round of research till the moment I worked on the final version of the manuscript. Also, thanks to all the family members and friends for their support and enthusiasm. I would not have become successful in preparing this book without their love and concern. Last but not the least, I would like to thank Bidur Dongol from the Vajra Publications Inc.Pvt.Ltd., editor Shiva Rijal for helping me publish this book.

Chapter II

Kike Means Khukuri

As citizens of a nation that is famous for its Gurkha warriors and their khukuris, it sounds quite awkward to say that we lack authentic written documents on the history of this very weapon. No matter how awkward it may sound, it is our responsibility to write its history now. Difficult may be the job, but a duty is a duty, and one needs to perform it sincerely. Scattered may be the facts and information about khukuri's history, but they have not gone into the oblivion. Nor have they been stolen away from the universe. It's now our job to synchronize such dispersed stories and facts and weave a new narrative. We can't blame our ancestor for not writing a proper history of this weapon. Let's realise a truth that they weren't fool. On the contrary, they were so wise that they regarded khukuris as integral part of their myth, rituals, daily life and political history of nation. Now it's our job to prove that we too are no less wise; otherwise our grand children will look down upon us for not recognizing the values of our cultural assets. It is paradox that often times we travel around the world searching meanings, and exasperated return we to find it at home to our joy and surprise.

One comes across significant historical facts that weapons similar to Nepali khukuri were in use in ancient times at different places. The Egyptians called their favorite weapon 'kopes'. Kopes were so magical that the Greeks were supposed to have copied its design. They called it 'kopis'. The Romans were so fascinated by it that they continued using this unique forward curved blade-shaped weapon and called it 'machira' that was called 'falcata' latter. This was the cavalry sword of the ancient Macedonians. One can conjecture that Alexander's horsemen could also have carried this very weapon with them in the 4th century BC, when they came to invade to the North West India. There is also much similarity between the manufacturing process of khukuris in Nepal and the sword-making procedure followed in Japan. As the finest sword in Japan is made from 100 layers of finest metal so is khukuri. Samurai sword-making technique in Japan has made its long history, and people and ideas have always traveled from all directions for ages. What one can say is that khukuri is an archetypal form of weapon. Varieties of khukuri like weapon seem to have appeared almost in every part of the world.

Khukuri is ancient weapon for Nepalis, especially for the Kirantis. Mention should be made here that the Kirantis, the Mongolian stock came to power in Nepal in the 7thcentury BC. The kike is known to be the soul of their culture. It is certainly one of the oldest traditional blade-form of weapons used by their ancestors, writes the Gurkha Khukuri House, one of the established khukuri manufacturers of the country in its website. Though kike has been in use since time immemorial, the Rajputs, the Kshetriyas from India are believed to have introduced a similar weapon known as khukuri now as they migrated to the region known today as Nepal. But there are no written documents and other forms of historical facts to prove this conjecture. Some also believe that khukuri might have originated from a form of knife first used by the Malla rulers who came to power in Nepal during the 13thcentury AD.

From the perspective of a retired British Gurkha and a Rai by birth and culture, I see strong connectivity between and among myths, legends, history and contemporary state of kike or khukuri. My ancestors were believed to have come from the east beyond the borders of the present day Nepal. By around 1500 BC, they began to establish and expand a kingdom starting from the far east of Nepal. By 600-700 BC, they reached the Kathmandu Valley. After the Ahbir or Abhir dynasty, the Kirantis invaded the country from the east and occupied the Valley. Their reign said to have lasted for 1000-1500 years. Their capital was believed to have been Matatirtha and Thankot, the west part of the

Valley. The Valley was once called 'Kiranta Desha' or the country of Kirant. It was one of the most fascinating countries lying in the extreme east of the Himalayan foothills. The three main hoards of the Kirants had invaded the Valley successively, and this event was supposed to have taken place towards the year 700 BC. The names of the kings appearing in the Vamsavali are 29 in number. They have a close affinity with the present day Kirantis i. e. Kulung, Thulung and Yellung.

Since ancient times, kike has remained Kirantis' choice of weapons. And, one can call it an early version of present day khukuri. At home in hills and mountains and when and wherever they move, they always carry it with them. The women carried their 'khurumbi' or so-called 'khurmi' from dawn to dusk.

During the ancient time, there must have been many skilled kike makers in the land of Kirantis. It is believed that some supreme kike masters must have set up foundations for close-quarter kike combat in the Himalayan regions. Every community has its own unique style. Each has distinct superstitions, rituals and spiritual practices linked with kikes. There are diverse tribal shamanic rituals and distinct khukuri practices performed by Rais, Limbus, Sunuwars, Gurungs, Magars, Lamas, and other tribes remain unique and diverse.

Listening to myths, legends and beliefs the Kirantis associate with kike, one can conjecture that kike has long history, going back to thousands of years. Ancient Rai shamans and gurus used kike for various rituals and ceremonies such as: spirit communion, exorcism rites for families, appeasement of gods, worshipping of weapons, practices of meditation that clears the clouded mind, performance of Maha Mudra (tantric ritual), healing rituals, casting off demons from villages, animal sacrifice, funeral and burial ceremony among other initiation rituals and ceremonies.

Rais believe that our ancestors are never dead until and unless we forget them. People, who never look back to their ancestors, will not look forward to posterity. Lord Veen shelters the Rais from any strangers, and gives them knowledge, strength, protections, and makes them fearless. Kike itself is a distinctive knife used by the Kirantis as a weapon and also as a spiritual object. It is often considered to have magical powers. It is both weapon and symbol, forged as a religious act and wielded with the super-human skill of Kiranteswor, the lord Shiva.

Figure No. 1: Sketch of Yalambar Hang, the first and great king of Kirantis

"Mera bir Kirantiharu Sher jasta chhan
Jahan pharmaunchhan, tyahan phatte garchhan"

(My brave Kiranti are lion like warriors
They win every battle they take part in)

Myths and legends about kike become much stronger as one visits the museum and observes the repertoire of kike like weapons used by people for ages. Kiranti's earlier kikes and their other most identical weapons are still displayed at the National Museum Chhauni (NMC), in Kathmandu. Such displayed weapons include:

1. Ancient kike in different versions or models.
2. Khurpa of different versions, 'khurumbi' carried by Kiranti women.
3. 'Chhuri', or 'Chakku', an old version of knife for stabbing and thrusting uses.
4. Kachiya or sickle with sharp shark like teeth used by women to cut grass, barley, wheat and rice and for other uses.
5. Bancharo or axe, very similar to tomahawk, which gives enemy maximum penetration. It weights 1 kg approximately. Its handle is made from rosewood. Kiranti warriors would carry it with them in the past.
6. Dhanush and karh or kadh/ bow and arrows. In earlier days, Kirantis were experts in archery also.
7. 'Thaankro', quiver made from bamboo bound with leather, poised arrowhead 2' 8" length.
8. 'Guleli' and 'matengra', bow and pellet. It has length of 5' 11". It is made of thin strips of bamboo. In the centre hangs a pocket to hold 'matengra' (clay balls) or any round pebble for hurling.
9. 'Ghuentro', a leather belt for hurling stones, the size of a cricket or tennis ball to be thrown at enemies. Also used for throwing of any written messages wrapped on paper, animal hide for troops.

Today, Kirantis are represented by Rais known as Khumbus, Yakhas, Limbus or Yakthumbas, Sunuwars, Tamangs, Sherpas, Lepchas, and Bhutiyas of Nepal, Darjeeling, Bhutan, Sikkim and Burma. Kiranti tribes have become so intermixed and have so many practices in common that they are now considered as one. Mechi River marks the frontier in between Sikkim, India and Nepal. Kirantis were first reported in Sikkim in 1812 by great Indian traveler Sarat Chandra Dass, when he visited Tibet via Sikkim.

Mr. Subash Ghising, a former leader of the Gurkha National Liberation Front (GNLF) from 1986, once said that 'The Khukuri is mightier than the pen'. 'Quitting is easy; fighting is hard. Quitting is losing, fighting is winning so 'Ayo Gorkhali!'.

'A Gurkha has a unique characteristic- in victory, he advances, in defeat, he charges. There is nothing like retreat in his psyche', wrote Lt. Col. Daljit Singh, an Indian Gurkha Army officer.

'The harder the battle, the sweeter the victory'. There is a wealth of truth in this ancient saying of ancestors of the Kirantis.

Kikes and Indigenous Nationalities

Today, many indigenous nationalities in Nepal believe that Kirantis were the original rulers of the land. And, they are still the best khukuribaaj of this mountainous country once known as 'Kirant bhoomi' or the land of Kirantis. Khukuri was and still is their 'birtako nishan' or symbol of bravery in Nepal. They have their own scripts, languages and words for Khukuri or kike and other accessories. Even today, there are 59 indigenous nationalities in Nepal and some of them do have their own word for kike or khukuri. I would like to provide the following chart to point out the rich heritage of weapons that the indigenous communities of Nepal have lived with for ages. This rich range of vocabularies makes us ponder over the homebrewed ages old talents and skills gone in the making of khukuri or kike.

Indigenous Communities	**Local names for their weapons**	**English**
1. Bantawa Rai	a. daabe	Kukri
	b. daabeship	Sheath
	c. bhokshi(chakmak) Seb Ban (Chakmak)	Blunt knife
	d. dakuku(karda) Chuk Ban (Karda)	Sharp knife
	e. khopiya (khisa)	Leather pouch
	f. kombi	Thick curved sickle
	g. vee(karh)	Arrow
	h. hekmu (kachiya)	Curved sickle with sharp teeth
	i. mejom(bhala)	Spear
	j. gure (matengra)	Clay balls
	k. kuupu (chupi)	Dagger
	l. wantuk (bancharo)	Axe

	m. aptusang (banduk)	Rifle, gun
	n. karukan (reti)	Iron file
	o. khorek (arii)	Hand saw
	p. sayaban / Tuliban (tarawar)	Sword
	t. Meera Lung (Dalchhin dhungo)	Quartz/pebble
	Chhapship	Pen
	Chhap lab	Paper
	Lungwain (ghuentro)	Pallet bow
2. Chamling Rai	a. khiipa (khukuri) b. talek (dhanush) c. bhee (karh) d. bhechuk (bhala)	Kukri /Khukuri Bow Arrow Spear
3. Thulung Rai	Kike, betho	Kukri /Khukuri
4. Kulung Rai	Kike, kheepa	Kukri /Khukuri
5.Sampang Rai	Hooga	Kukri /Khukuri
6.Yallung Rai	Kike	Kukri /Khukuri
7. Limbus	Pheja	Kukri /Khukuri
8. Magar	khorchey, heja khorchayang, khorchayar chieun	Kukri/khukuri Kukri/khukuri Kukri/khukuri
9. Gurung (Tamu)	Kohoch	Kukri/khukuri
10. Sunuwar	chuub khoorchi phunglu	Kukri/khukuri sheath Quartz to produce fire
11. Sherpa	khurpa thieun meecha dosertuk raah (jhulo)	Kukri/khukuri sheath Blunt knife… sharpening Khukuris blade Quartz to produce fire, Fire tender
12. Tamang	Gojaa	Kukri/khukuri
13. Jirel	thieun thieun shup (dap) thingma (karda) mejaak (chakmak) sherdakdo seewa (jhulo)	Kukri/khukuri Sheath Sharp knife Blunt knife Quartz to produce fire Fire tender
14. Rajbangshi	khukuri dhaula pathal (jhulo)	Kukri/khukuri Quartz to produce fire

Kike

Before I move further, let me introduce kike in brief. Culturally, I am comfortable describing it because I feel connected to it. Kirantis, the community I belong to, have given it a sacred stature for ages. For Kirantis, kike's size and shape has symbolic meaning. Each contour of it has cultural and spiritual value. Every Kiranti guardian tells his or her children that kike's shape and its parts have distinctly cultural values and meanings associated with. The outer shape of the blade itself looks like a crescent moon. I also regard it as a symbol of Nepal. And the notch near the hilt has a trident of the Lord Veen, the God of destruction and war. I must say that a mere look at kike makes Kirantis become spiritually and emotionally larger. Kike is not a mere weapon; it is the way Kirantis look at their life and the world around.

There is a famous riddle among Nepalis. It goes: what is that which lives inside a black house and drinks red water? And the answer is: Khukuri. Its 'kshura' or razor sharp blade speaks its own language. For early warning, it gives a straight quick sharp chop as it comes out of its sheath, and then slash and thrust as a reply to any challengers. 'The boss is not always right but the boss is always the boss'. You can say exactly the same thing about khukuri as well.

The fact that kike's each part is significant and connected to the whole tells us about the volume of creativity that might have gone in its making since ancient times. For example, its metallic round shaped plate between the blade and the handle is called kanjo or bolster. With its curvature shape kike weights less. Its contour is called chira or fuller. Its notch is called kaura(ri), meaning a small seashell. Significantly, the 'kaura' in Kiranti Rai culture indicates 'ragat bagne dhara' or a blood gutter. Kirantis use 16 pieces of 'kauras' when we sit to play gambling mainly during the Deepawali festival. There are dozens of distinctly named parts of khukuri, and each of them has its own significance. Khukuri, therefore, is a sophisticated weapon.

I remember my grand father telling us many dramatic stories about fortunes lost and won in the villages during the Deepawali festival. The gamblers would go so much crazy that some of them would loose

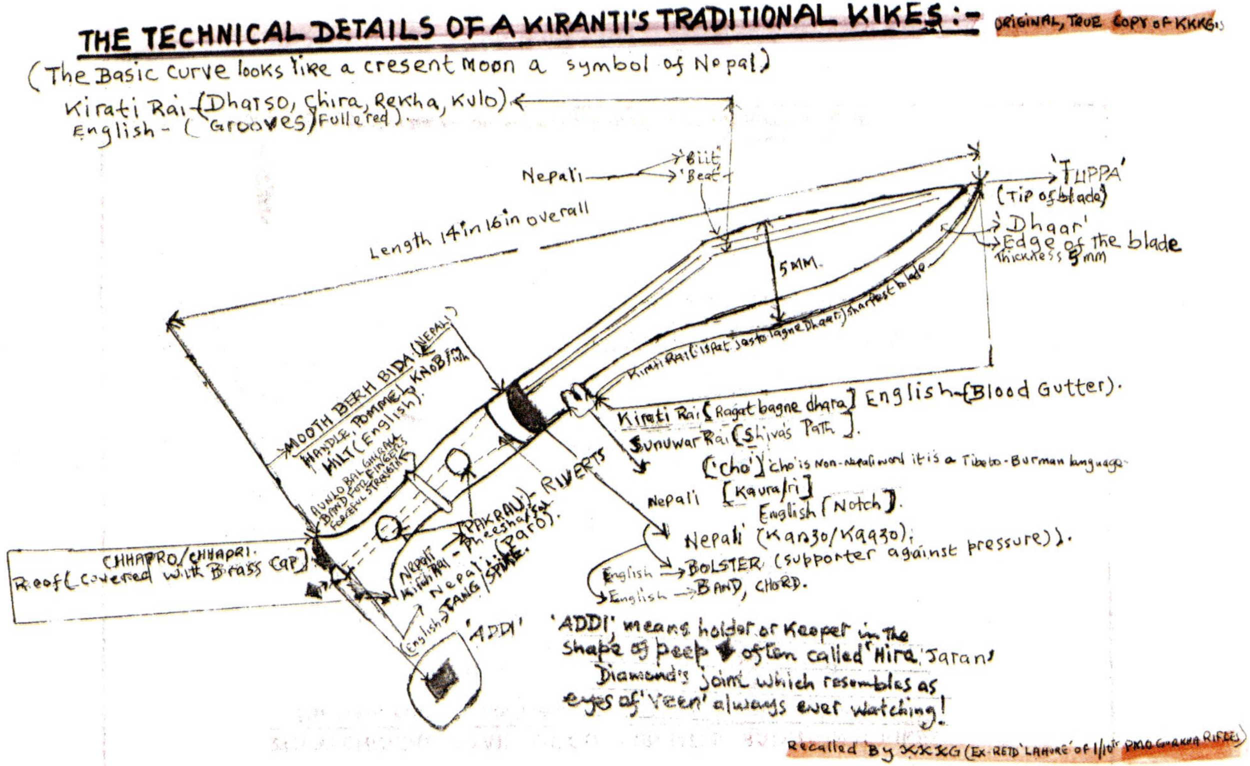

Figure No 2: A sketch of Kike (Source: KKKG)

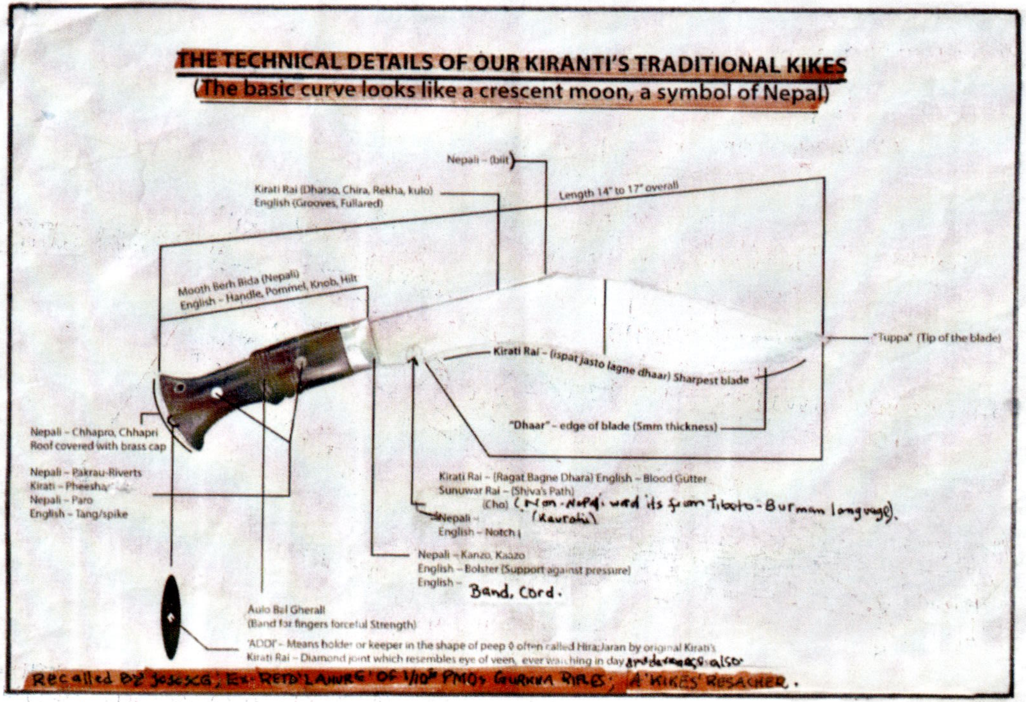

Figure No. 3: Sketch of Khukuri (Source: KKKG)

animals, livestock, a complete farm and so on. So much depended on sixteen pieces of kauras. Today, it reminds me my own regiment. During the Deepawali, every issued and private khukuri would be put under lock and key in each company's arms kote or storehouse until the 'juwa' or gambling would be called off. At home Kirantis keep three kauris into their '*khisha*' or small pouch they carry for some tantric reasons.

The "kauri/kaura" or notch in the blade near the hilt of khukuri arouses much interest. Some regard it is a Hindu religious symbol representing a cow's hoof. Some take it as an air channel for easy extraction from the wound. Others suggest that it is like a rifle's foresight meant for aiming the target. Whatever may be the meaning, we have to realise a fact that khukuri is more than a weapon.

A Gurkhali never ever throws his khukuri; he always prefers to keep it under own possession with pride to his last breath. The reasons are:

A. During any close quarter combat, khukuri can neutralize the power of opponents' khukuri, sword, 'roal' dagger, and 'katari'. It can offset enemy's sword and bayonets in the war.
B. In close quarter combat, it can act as a blood gutter, and can check excessive blood on the pommel.
C. In earlier days, Kiranti men used to carry an extra axe to throw it on any running away targets at lethal range up to 12 feets. The samples are still on display at the Nepal Museum Chhauni (NMC) Kathmandu.

I would like to introduce some Nepali terms used for each part of khukuri simply to evoke its sophisticated nature:

Khukuri ko dhaad: Khukuri's spine, back blade.
Khukuri ko dhaar: Khukuri's main blade.
Khukuri ko tuppo/chucho: Tip of khukuri.
Mooth, Bida, Berha: The handle, pommel, knob or hilt.
Aunla bal: The circle in the middle of the handle made to enable fingers to clutch it firmly.
Pakrau, Phesha/so: Rivets.

Chhapro/Chhapri: Roof or brass butt plate, cap.
Addi/Phuli/Hira Jaran: Holder or keeper.
Karda: A small sharp knife for skinning and carving uses.
Chakmak: A small thick blunt knife for honing blade and to ignite or produce fire.
Kalam: A pen (originally made from 'malingo' or bamboo).
Khisha: A leather pouch used for keeping khukuri or kike. This leather pouch generally holds the following essential items:

a. Kagaj: Nepali handmade papers to write letters or notes.
b. Jhulo: Fire tender, prepared from Nepali paper, cotton, wool, banana stem, brown polypore, asteraceae ('Titey Paati'), 'bukki' and shiru flower etc.
c. Dalchhin Dhungo: This flattened and steamed polished white pebbles were probably one of the most widely used devices to spark fire in the early phase of human civilisation. 'Dalchhin Dhungo' is struck against a small steel bar called 'chakmak' to produce sparks which set on fire on 'jhulo'.

In the high rugged Nepal's Himalayan region, the weather remains always misty, cold and damp. In such condition, the Kirantis of ancient times, one can conjecture, might have taken 'dalchhin dhungo', 'chakmak' and 'jhulo' as essential goods to run life. Such items were in fact the heartbeat of human life in this corner of the earth. Even to this day, sparking fire in medium other than with 'dalchhin dhungo' is regarded as going against the Chomolungma (local name for the Mount Everest) or the law of the local land.

It is obvious that modern medicine and doctors aren't available in the remote mountains of Nepal. Most of the highlanders and herdsmen in Nepal have good knowledge about medicinally significant herbal plants available. Many herbs can be used to cure one from venomous snakebites, scorpion stings, wounds, fever, rheumatism, asthma, skin diseases etc. The herdsmen in the mountainous jungle always keep important herbs in their khisha such as *silajit*, *bikhuma*, *hing*, *chiraito*, *bozo*, *yarsagumba*, *jhaar mauro*, wax etc., some of the famous herbs among locals.

Pommel or Handle

Khukuri's pommel or handle is known as 'mooth' or 'berha' or 'bind' in local dialects. Its handle is usually made from locally available hardest woods such as walnut, rosewood, sissoo (North Indian rosewood), Indian sandalwood, and rhododendron wood among others. The expensive 'bind' is usually made from water buffalo's horn, deer horn, ivory, aluminum, white metal brass etc.

The Khukuri, Nepal's Royal family members would carry in the past, looked distinct, as its pommel or bind would usually be made of ivory and rhinoceros' horn. It would frequently be decorated with two or more precious gems, usually emeralds.

The 'laha' as strong as today's epoxies is formed out of the mixture of pitch, honey and other saps, and then poured hot into the carved out pommel so that it can hold 'kanjo' and 'chhapro' tightly.

'Pakrau' or Grips

Rivets are used to secure grip in khukuri. Some also use flat long nail or 'paro' locking itself round. Sometimes, the grip is secured by means of screw sunk into the pakrau. In some old examples, the brass butt plate or cap is riveted, giving impression as if two odd dots were engraved. This is mostly found on earlier versions of khukuri.

Sheath (Khol or Daap)

The sheath for khukuri is made of two pieces of plain wood frames. Frames carved from some rosewood or rhododendron become perfect material to provide it some artistically appealing looks. Sheath made of this type of wood material look polished and strikingly beautiful. Chainpur of Sankhuwasabha, a district in the eastern part of Nepal, is famous for such woods. Commonly, the sheath is covered in the skin of goat, sheep, yak, buffalos, rhino and even elephant. Some good quality sheaths are made from aluminum, white metal, brass and copper. Some are very ornamentally embellished with brass, filigree work and heavily set with colored glass, quartz and even beryl.

As mentioned earlier, one can easily recognize the sheath of khukuri the royalty and distinguished personals carry as it usually is decorated with gold or silver or any other precious gems like aquamarine, turquoise, lapis lazuli, coral etc.

Namja or Farash for Sheath

Namja, the leather belt used to carry sheath, was supposed to have been introduced by the herdsmen of the Himalayan jungles. Later a modified version of carrying khukuri in 'sash' method also came in practice. The 'namja' is often made from raw nettle skin and animal hides that is separately tied round the waist to hold the scabbard.

During the Malaya emergency, the Borneo confrontation and in IS situations in Singapore and Hong Kong, the Gorkhas carried their 'daabe' in a Farash attached to the waist belt next to ammunition pouch or often behind the waistline. Khukuri scabbards were covered with OG clothes to match the uniforms and equipment. The practice of covering the scabbard of khukuri and sometimes the bayonet with khaki cloth appeared in 1929.

It is not known when the 'namja' was first introduced in the Gurkha army, but a photograph in the Gurkha Museum of the Nusseree Battalion taken at Fattegarh in 1858 shows a khukuri scabbard being held in wide frog or an attachment belt. Today, the British Gurkha soldiers carry their ceremonial khukuris for both guard duties and ceremonial parades in a leather-laced black 'namja' or frog.

Maintenance of Khukuri

Like all other arms and weapons, the durability of khukuri depends on how we take care of it. After every use, it is recommended that we should dirt-free it with a piece of clean dry clothe before we tuck it into the sheath.

Remove the Rust

Chakmak, sand ash and charcoal are the best means to remove rust in khukuri. If available, it is better to use steel wool, sand paper, and emery cloth soaked in a few drops of kerosene or petrol. Water and moisture can cause the iron catch rust. Therefore, my advice is to dry your khukuri before you tuck it into the sheath.

Sharpen the Blade

Nepalis use 'chakmak', a small size knife to sharpen the mother khukuri. During the World War I and II, the British Gurkha soldiers used stone, even bayonet and steel helmets to sharpen their khukuris. A tungsten carbide sharpener can carve a razor edge on khukuri, machete and other types of knives and cleavers in an average of ten seconds.

Re-sheathing

Once the rust is removed and the blade tested and adjusted, it is time to polish khukuri with oil before it is re-sheathed. In the Himalayan region, the herdsmen use animal's fat, ghee or cooking oil to polish it. At any case, hair oil or cream meant for healthy looking hair can also be used to polish khukuri as well.

It is best to grease all metal parts of khukuri regularly. Chemical like metal glow, the aluminum oxide polishing paste can also be used to clean, polish and protect khukuri. All leather parts of khukuri such as scabbard, 'namja', and khisha can be cleaned and polished with boot polish.

Homemade Recipes

My grandfather, Shree Bhairav Sing Nechali Rai, understood khukuri very well. I grew up watching him cleaning his khukuri at home with love and passion. The materials he used to clean his khukuris were simple and easily available at home. It is best to give these homemade cleaners another try! His ways are safer and cheaper too. At times when the bottled and packaged chemicals are loathing us as well as harming our environment, and producing toxic fumes and hazardous waste, it is best to turn towards my grandfather's recipe for cleaning khukuris and other metallic pans and pots:

1. To clean copper vessels, he used turmeric and lemon. The result was fantastic. The vessels would gleam smilingly.
2. For brass vessels, he used a 50/50 mixture of salt and flour with a little vinegar added. And, the result was spotless, bright and shinning brass vessels.
3. For silver pots, my grandfather would, first of all, boil one-liter water with a tablespoon salt and baking soda for three minutes. He would drop semi thick liquid on silver vessels, and then polish them with a soft cloth. The pots and pans would shine like the moon.

Testing Blades

Experts in the past could make blades of crudest variety. For example, they used 'paain', the most sophisticated-watered steel. Sharpness of blade depended on interest of the person who places the order. Much also depends on the authenticity of steel. Khukuris made from inferior quality of steel can't hold sharp edge. Whether khukuri's edge is sharp or not, one can easily identify it through following procedures:

1. For on-the-spot test, strike one khukuri's blade against other. The khukuri that makes the slight nick is of good quality.
2. Place your thumb's nail on the edge of blade and try to press downward from one end to the other. The good quality khukuri's 'tick' sound remains more or less the same.
3. The best method is to use 'chakmak'. With it you beat on edge from one end to the other. The good quality khukuri produces sound of the same beat.
4. The final test is what locals call the 'lathro' practice. You can strike khukuri on any hard target like buffalos forelimbs or on any hard timber to guarantee its 'paain' or edge.

5. The best thing is to have a guru by your side. Once demonstrated by your guru, you need not follow all these procedures or else it would be like taking medicine brought without prescriptions.

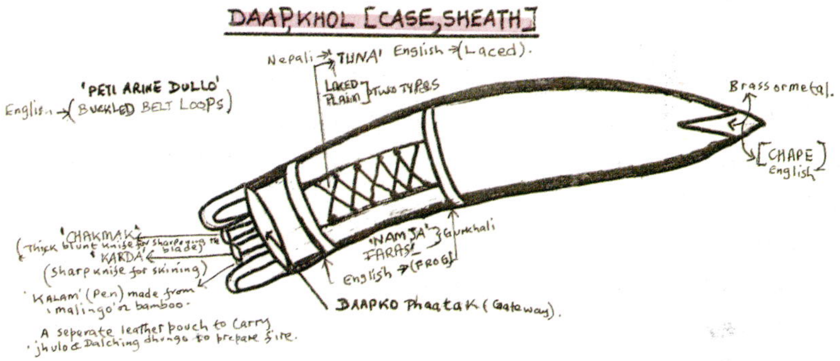

A.

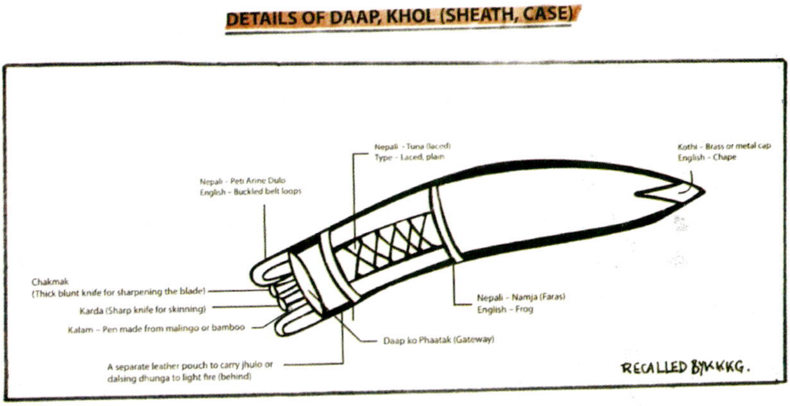

B.

Figure No. 4A/B: Daap, khol, case or sheath (Source KKKG)

Kike and Geocultural Reality

In the high Himalayan region, herdsmen usually are found playing some rustic games like lathro or dandi biyo, westerners can call it a game of stick. Another popular game among them is khopi or a game of coins. They also play music to pass their time and to entertain each other. They play murchunga or jaw harp, binayo or the bamboo harp, murali or flute, madal or drum. Importantly, they sing folk songs not only to entertain but also to express their inner life.

In the forest where wood and bamboos are plentiful, all herdsmen test their kike by chopping green trees. Kike they hold usually has the size of 3 to 4 inches in diameter. It is big and sharp enough to chop branches at one go. How one is going to use kike depends on the bent of branches. Branches are either straight or can have semi-straight bent or some can have complete bent. While chopping such branches herdsmen use or follow three paths of kike: straight, horizontal and diagonal. This practice of chopping branches and plants of all dimensions is very important for conditioning thumb, fingers, wrist and balancing kike. One also happens to train to chop soft as well as hard targets in this process. Experts will chop in one blow the size of a man's leg or neck size tree or bamboo with ease. Some practices are carried out in a stationary pose. Some practices are done with kneeling. And, some practices are done running and jumping with kike.

It is said that one morning Sir Winston Churchill was sitting in his bathtub and enjoying his time. His servant next door heard his master's voice. The servant is supposed to have opened the door and asked, "Were you speaking to me, Sir?"

"No, I was addressing the House of Commons", replied Churchill annoyed at the interruption.

In the same manner, Rais feel annoyed when they are disturbed while working with kike. Kirantis practice their kike seriously. They use it for all kinds of purpose.

My grandfather brought me up at Ilam, the eastern Nepal. He used to spend most of his days in mountain looking after his over 100 cattle. He was a well- reputed 'sardar' or merchant for his butter, cheese, milk, and curd business in Ilam bazaar. Every Wednesday and Saturday, he used to supply over 60 'dhakars', bamboo basket carried on back, loaded with above-mentioned goods to the wholesale buyers.

I remember that most herdsmen had their own styles of carrying khukuri tucked on their 'patuka' or cloth belt. Some would also sling it in sash method. Most of them carried 'budune' type of khukuri, and few carried 'sirupate'. My grandfather carried his favorite 'ganjewal', always tucked in his red 'patuka' from dawn to dusk, and used to place it under his pillow whilst sleeping. I saw that he would use this khukuri to chop down thick, long and heavy wood from jungle, as he needed to build cowsheds. For such hard works, the short, light and ordinary khukuris aren't of much help. Sometime, I used to mess around with my grandfather's khukuri that was too heavy for me to lift and use it. Surprisingly, one day, he brought a 10-inch long sirupate khukuri for me. Before handing it to me, his first advice was "don't you ever draw your khukuri unless required for 'pranayama' or prayer, chopping wood, clearing of shrubs and severing chickens". Indeed, I was so pleased to receive my first and memorable khukuri from his hand. In no time, I understood the reason why he discouraged me from drawing khukuri without any proper reason. One needs to avoid unnecessary accidents from the blade.

Phaudari or Combat

For Kirantis, phaudari brought opportunity to learn and demonstrate skills and techniques of using sirupate khukuri and roal. The Rais and the Limbus in earlier days would organise such phaudari or competition that brought khukuribaaj to their doors. Such game would bring fighters or kikebaaj from villages and neighboring districts around. According to the olden customs and rules of such combat, no complaints would be accepted for 2– 4 inches deep cuts, loss of arms and legs. It is rumored that fighters being separated into head and body would be the subjects of consideration in such combat. Jokes and facts about the combat of Rais and Limbus abound. This could be one of the reasons why the famous British Field Marshal Sir William Slim once said: "For the Rais and the Limbus, I think, it is to excel in marlin dogfight of close-quarter hand to hand combat".

Phaudari Preparations

First of all, some man-sized dummies are prepared from thick tree barks, dry grass, and hay or straw. They are tightly wrapped on a wooden or bamboo pole. Gurus usually point out seven important target zones on dummy as well as on the body of kike practitioner. Then they demonstrate about how to carry out the CTS operation on target object depending on whether they are in the state of stationary or motion.

Before any practical stage begins, a guru demonstrates the ways kike can be used step by step. He is supposed to teach the learner disciple to develop his real skills, techniques and talents so that he inculcates capacity and talents to live and fight until his last breath in battle.

Kike gurus train their students mentally and physically for months. The result is a brave warrior who is able to face all kinds of encounters with talents and techniques fearlessly.

Basic Training for Kike Combat

1. Running in full speed, and cutting the target.
2. Jumping into foxholes and bunkers, and cutting the target.
3. Crawling on stomach, and cutting the target from the prone position.
4. Dropping and rolling on the ground, and cutting the target.
5. Squatting and kneeling, and cutting the target.
6. Carrying a wounded person or heavy load, and cutting the target.
7. Blocking and parrying against bayonet or sword of the target.
8. Jamming against rifle-butt, and cutting the target.
9. Multi directional cutting.

Functions of Kike

During the war, kike serves three major functions. Based on such functions, one can put kikes into three types: combat kike, ceremonial kike and communion kike.

1. Combat kikes are used for close quarter, hand-to-hand fighting in land, mountains and jungle warfare.
2. Ceremonial kikes are used for royal palace guard duties, governors' house guard duties and visits of dignitaries. It is used for street linking for VIPs, regimental and ceremonial parades, regimental khukuri drill competition, regimental guard duties, kikes' inspections, to guard defaulters, and for funeral ceremonies as well.
3. Kiranti Rais take their communion kikes including Sayaban (swords), dhanuskarh (bows), bhala (spears), aptusang (rifle guns), khurumbis (sickle) and axes in union with souls of their ancestors. Such kikes and weapons are supposed to passess warriors' spirits in them. This is the reason why they are secretly kept in their prayer room. Only the senior elderly members of the families are allowed to enter the sacred room allocated for the ancestors and the family deities. Culturally named as *Mang Kotha*, it is where worshiping and poojas take place. The Kirantis have continued this tradition ever since time immemorial.

Effectiveness of Kike

The devastating effects of this weapon are as follow:

1. The neck is the best target zone for a clean chop. A fine blow on this spot can paralyze arms for sure. Severing the spinal chord is usually the most reliable way to stop any determined attacker. The lower cut and thrust will cause the opponent to fall down, thus making him an easier target.
2. Smashing the skull if done with a heavy blow can usually stun opponent if not completely stop him. Smashing bones like hand and feet and disabling opponents can be a smart tactic. Kike's curved blade excels for such dismemberment.
3. Serving or breaking shoulder bones and limbs is also a good way to stop or at least spoil any opponent's energy. A wounded limb is enough to stop even the toughest, most determined opponent. Even if he does not stop, loss of blood will definitely lower blood pressure

within 3-5 minutes so that brain will not receive further oxygen, and opponent will rest in peace forever.

4. Making deep wound by cutting, thrusting and slashing will result in rapid blood loss. Thrusting deeply into vital organs will also cause severe blood loss and damage the energy of opponents.
5. Thrusting and slashing major veins and arteries with sirupate khukuri is the best one can do during the encounter. All other heavy knives do well also. However, one of the most deadly moves in any knife fight is the reverse cut; this is where long heavy budune or bowie knives can dominate. A reverse move can also be done with kike, but because of its dull dhaar or spine, it will only break or smash bones; whereas today's modern K3 double bladed kike can do this job better.

Soft and Hard Targets

Cutting, thrusting and slashing (CTS) with accuracy and power against both soft and hard targets is not an easy task. In contemporary warfare, troops often have to breakthrough enemies' defensive positions by destroying all kinds of hurdles. They use wire cutters or khukuris to cut low and high barbed wired obstacles. Needless to say, one uses demolition or explosives. In some cases, one can use khukuris to cut hard fence posts, electric and telephone's wooden poles.

It is possible that khukuri sometimes flies out of hand as one sets to hit hard targets. One should realise that securing this weapon in hand requires strong fingers and wrist, especially to cut things like truck's tire tied on fence posts, telephone poles, swinging wooden blocks. One should use it very carefully to parry helmets, shovels, picks, short iron piquet etc. thrown at him. One needs to maintain mental focus, concentrate on strong breath pattern and crawl with correct body postures and use khukuris to cut the soft targets.

I should say that drawing and re-sheathing khukuris while walking, running, crouching, CTS with speed and accuracy against any soft or hard target is again not that easy task.

The grip achieved for holding khukuris varies from left to right hand. One should grip it differently while using both hands to hold khukuri. The grip one uses for holding khukuri is definitely different from the one that is used to hold dagger or kattari. Moreover, whether to tighten or loosen the grip depends on circumstance met on time.

Soldiers in training need to go through daily rigorous exercises and practices continuously. Holding khukuris all the time cause one live with plenty of sweats, tears and even blood sometimes. Their one and only concrete mission in mind should be that "Now, I am ready and I can do it".

Any real khukuribaaj can draw both his khukuri and his roal within a second. Khukuri combat tends to be very fast. The three fastest and accurate actions one should perform at one go is 'chhyaat', 'chhyaat', 'chhyaat', or what I call is the zulu 'z' strikes.

Attacking and charging with drawn khukuri is extremely dangerous, and is very difficult to observe safety principles. This is why our ancestors and forefathers used the Rai Phaaro's technique. This means Rais should not bother about the identity of opponent, foe or friend whoever comes on sight, they should perform their CTS techniques precisely.

Khukuri's Charge

In battlefield, khukuri-charge is carried out in the following conditions:

1. When group attack is launched in and along foxholes, trenches and bunkers.
2. When group attack takes place against enemy's outpost and patrols, and also during enemy's resting, bathing and feeding time (hours).
3. When one sets to attack at night to terrorize and demoralize the enemy.
4. Makeshift enemy bunkers, foxholes and trenches are extremely dangerous places for one to attack with khukuri. One finds it very difficult to observe the safety principles in such spaces or conditions. Moreover, one has to follow the famous strategy "No friend, no foe".

A khukuribaaj moves with the CTS principles, i.e. cut, thrash and slash. For example, during the Sino-Indian war in 1962, one Gurkha soldier is supposed to have killed repeatedly nine enemy soldiers with his single khukuri before killing himself. In the age of machine guns and automatic rifles, Colonel Lewis Millet is said to have led the last major bayonet charge by the U.S. soldiers on Feb. 7, 1951 attacking a group of Chinese and North Korean soldiers with bayonet on a hill near Osan, South Korea.

Multi-functions of Combat Khukuri

1. Chopping

This is the name given to the action when one uses sharp blade of khukuri to chop neck, wrist, arms, spine, hip or leg.

2. Thrusting

This is the term given to the action when one uses tip of blade to thrust at eyes, throat, armpit, chest, belly, ribs, hips or groin area.

3. Slashing

As part of this game, one uses lip of blade to slash at face, neck, throat, joints or arms and legs of opponent.

4. Slapping

As part of this strategy, one uses flat of blade not to kill but to stun enemy by slapping on face, skull, neck, shoulder, knee and elbow joint. This method is found to be very effective during interrogation of enemy prisoners and local native collaborators, informers and spies.

5. Hammering

One also uses back part of blade not to cut but to break bones, such as arms, shoulder blade, knee bone, shinbone, or hip-bone. This method is frequently used to hit enemy and not to damage blade.

6. Butting

One uses base of handle of blade against neck, face, or spine of enemy. This strategy is used when we get tangled with enemies in tight foxholes or trenches.

7. Hooking

One can also use curve of blade against enemy's throat, face, ribs, groin or armpit.

8. Spiking

One may use turned blade downward while facing any crouching enemy on elbow in battlefield especially in foxholes, between boulders and tree trunks.

9. Throwing

Throwing your khukuri at a short distance if enemy turns to run or escape can be its good use. I personally discourage soldiers from doing it. Instead of khukuris, you can throw axe and hit the person dead. Our ancestors would do the same for a short distance target.

Basic Understanding about Kike

Draw safely, hold safely and return it safely.

a. Overhand hold: One should point forward and the blade downward.
b. Underhand hold: One should point forward and the blade upward.
c. Sidehand hold: One should point backward and the blade downward.

Cleaning and Sharpening Kike

a. Use 'chakmak' to wipe out its rust.
b. Use sharpening stone to sharpen its edge.
c. If you are in war, use steel helmet, steel bayonet to sharpen its edge.

Safety Principles

a. Kike must be safely handled. One should not cut oneself.
b. It must not harm any innocent beings. One should mind the distance between kike and persons around.
c. One has to take care of its overall health.

Target Zones

a. Three high targets: center, left and right.
b. Three middle targets: center, left and right.
c. Three low targets: center, left and right.

Safety Zones (to avoid accidental cuts and control of weapon)

a. Above head
b. Left side of body
c. Right side of body

Levels of Targets

a. Head region (high level)
b. Body region (middle level)
c. Hip and below (lower level)

Paths of Combat Kike

a. Diagonal cuts
b. Horizontal cuts
c. Straight cuts and thrusts

Hand positions for Cutting or Thrusting

a. Lock-wrist for chopping and cutting hard targets
b. Snap-wrist for quick slashing and cutting soft targets
c. Twist-wrist for thrusting into internal organs

Types of Cuts, Thrusts, Slash (CTS)

a. Stationary CTS (from standing position)
b. Kneeling CTS (ducking to kneeling position)
c. Running CTS (running or sprinting after enemy)
d. Jumping CTS (jumping into fox holes, trenches)
e. Rolling CTS (avoiding spear, 'Kattari', 'Tarawar', 'Sangin')
f. Crawling CTS (moving stealthily at night)

Types of Defense

a. Blocking (against rifle-butt, sword, knife, etc.)
b. Parrying (against bayonet, dagger, sword etc.)
c. Jamming (against rifle-butt, baton, helmet etc.)

Effectiveness of Kike during Combat

a. At night, kike can be drawn safely and quickly in the dark and put to use effectively.
b. In jungle, places of many thick bushes, large and tall trees, low hanging branches, fallen logs, deep ravines, small and large rock formations, kike can be drawn safely and fast, and put into use effectively.
c. In trenches and bunkers, there is limited room to maneuver. Long tarawar and sangini are less effective in narrow spaces. But kike can be used effectively in such closed environment.

Kike's Defensive Moves

a. High right and left blocks (against swords, knife, bayonet attack)
b. Middle right and left blocks (against sword, knife, or bayonet cuts)
c. Low right and left blocks (against sword, knife or bayonet cuts)
d. Centre right and left parries (against, sword, bayonet, knife, or spear thrust)
e. Centre-line check (check straight weapon attack)

Khukuri and Designs

Bhojpur in the east and Pyuthan in the west are famous for manufacturing khukuris in Nepal. Designs of khukuris vary from place to place. For example, the long slender or shirupate blade is the characteristic of khukuri designed in 'Wallo' (this side), 'Majha' (mid side) and 'Pallo' (the far side) of Kirant regions. The shorter, round-bellied is often called 'buduna'. This fish shaped khukuri is commonly used in the western districts of Nepal. Similarly, khukuris of present times look different from the ones manufactured earlier.

There are many designs, styles and names given to khukuris in and around all 77 districts of Nepal. This makes one to realise that there are no specific sets of dimension of Nepali khukuri. One thing common among all kinds of khukuris has to do with the physique of Nepalis. Like most Asians, Nepalese have small but strong hands. The standard length for service and close-quarter combat khukuri including its pommel's 'chhapri' must be from 14 to 16 inches. Some earlier khukuris or kikes show widely differing curves on cutting edge. Observing closely, one encounters a line between the curve of cutting edge and that of back. The line varies from almost paralleled to that of a semi-circle, and the curve of the back itself varies considerably. One constant feature among khukuris is that all are unmistakably kike-shaped. Its blade is rarely decorated as it is made for close-quarter fighting weapon in battles.

However, one comes across commercially made khukuris in market these days. Such khukuris are decorated with dragon, yeti and temples.

They can be seen in most of the shops and streets of Kathmandu, Nepal. They are popular among tourists.

Following are the varieties of khukuris, which have remained in practice for ages. Since they have historical and cultural significance, I would like to call them kike or khukuri without much difference:

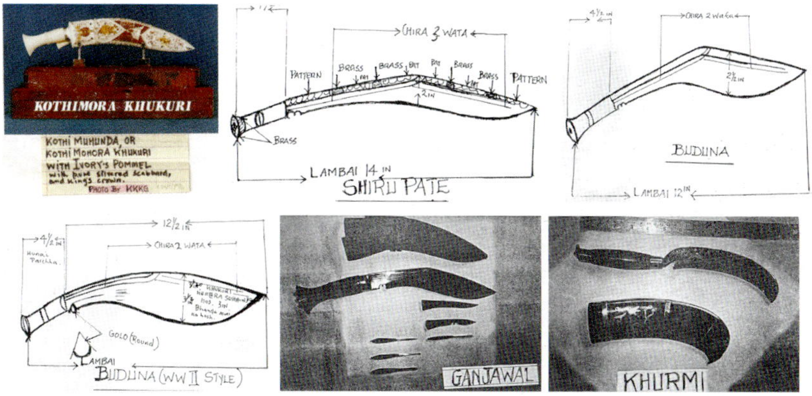

Figure No. 5: Sketches or photos of khukuris and khurmi discussed below

Kothimora
Sirupate
Bhojpure
Budune
Ganjewal Kukri
Khurmi

Kothi Muhunda Kike

Kothi Muhunda or Mohora kike is specially made from highly graded carbon steel. 'Kothi' in Nepali means a chain of joints. It is available in reasonable sizes though the military Gurkha issued-kike has distinct size. The gold-bladed, 15-inch-long Royal Kothi Muhunda kike of late king Birendra Bir Bikram Shah looks magnificent. At one time, it was placed on his royal throne during his absence from the Palace. Its sheath was covered with green, blue and red velvet on its top part. It was artistically hand crafted with peacock head for the crown. The pommel was made from ivory.

My impression is that Kothi Muhunda kike is especially made for gifts, presentation and memento for people who are to be honored. In the British Gurkha Army, it is presented to any retiring senior high-ranking officers of the Brigade.

Shirupate Kike

The long slender bladed, shirupate is the characteristic of early design of kike used by Kirantis of Nepal. Its original design is like that of 'shiru', a sharp and thin leafed-grass. It is made from high graded steel. Its sheath is originally made from rosewood, and is artistically hand crafted with one or extra sheath supporters. Its pommel is often made from brass, horn or aluminum. It is usually available in 14 to 18 inches size. It is the best ever khukuri for 'phaudaari' or close combat.

Bhojpure Kike

Bhojpur is a district in the eastern zone of Nepal. Khukuri made in Bhojpur is known as Bhojpure kike. Most skillful blacksmiths are still found in Bhojpur. With ages old art and wisdom at hand, they make religiously and traditionally significant kikes of the best kind. There is no provision for special stamp or mark of Bhojpur production. I strongly believe that there should be an officially recognized stamp for the Bhojpur-produced kikes meant for selling abroad. Chainpur of Sankhuwasabha and Dhankuta districts is also the best khukuri-manufacturing place in the eastern Nepal. The standard length and weight of Bhojpure kikes varies per the choice of buyers or anyone who places the order.

Budune or Buduna Kike

The design and shape of this particular kike is that of a round-bellied fish. People of Bhaktapur district of the Kathmandu Valley and Dudh

Koshi area as well as mainly from Kirant region commonly use it. It can have length between 11 to 14 inches. Its pommel is made from any hard rosewood or buffalo's horn. Its sheath is covered with buffalo or yak's hides. Its weight depends upon the size one orders.

During the WW II, the Indian Gurkha Army manufactured a ditto copy of budune kike. It was duplicated in Dehradun and was used in the World War II. The budune used during the wartime was an excellent close-quarter fighting weapon, especially for fast chopping of enemy sentry and any sitting or stationary targets.

Ganjewal Kike

In earlier times, herdsmen or yak-men from Solukhumbu and Sankhuwasabha district, the Himalayan range of eastern Nepal used this particular model of kike. Elderly people of this area still believe that 'Gyanje' is a name of a famous Sherpa of the ancient days, and he used to live in the high rugged mountains with his herds.

Gyanje, Dorje, Namche, Tawoche, and Neptse are very familiar words that one often hears in the eastern Himalaya. It mostly concerns with names of places, people, mountains, names of gods etc.

Among other kikes, Ganjewal has its own original characteristics, styles and traditions. It has its own traditional khisha that is mostly used for carrying tools such as 'jhulo', 'dalchin dhungo' and some folk medicines. Always at side, they can be of great service in any hard time.

'Ganjewal' can cleanly serve any Himalayan goats, sheep or even the neck of Yak in one clean blow. Its sheath is made of Yak-hides, and extra pattern of brown leather is sewn on top of black hide. It looks extraordinary in style. Its average length varies from 12 to 16 inches.

Khurmi

People call it khurumbi. Some name it 'komb'. Many also call 'baru' as it is short and easier to pronounce. Khurmi is a curved sickle and often called ladies' khukuri. Mostly, Kiranti women from the eastern part of Nepal carry it. Popularly known as *hansiya*, it is always carried on a patuka or clothes belt the women put. In one sense, it is also a weapon of all-purposes; at home, in the hills, used for cutting wood, clearing undergrowth and in any other chores.

Khurmi has the crescent moon like shape. Instead of fullers, khurmi has scroll patterns. It also has 'kauri' that is represented by a simple sharp-pointed edge near 'kanjo'. Another separate 'kanjo' is placed at the end of pommel. Instead of 'chhapri' that one finds in khukuri, '*hansiya*' has '*bulaki*' shaped metal ornament hung on it. This ornament symbolizes women's religious mark. This '*bulaki*' itself makes an odd and unusual sound as women carrying *hansiya* walk through jungle, and this musical sound is supposed to frighten off snakes and animals.

Khurmi or kombi is ritually treated on marriage ceremony in Kirant region. Mostly, it is used for communion services. In earlier days, it was called 'boju' khukuri also. Its metal, sheath and pommel are made from the same material as khukuris are made. The standard length of *khurmi* is 10 to 14 inches. Its weight varies per the interest of people who place their order. The same old '*chakmak*' is used for honing sickle's blade, which is safely stored in a lady's small bag '*thaili*' with other essential items. Kiranti women and '*kombi*' are often taken as inseparable in reputation.

Figure No. 6: Picture of Kiranti woman with a "khurumbi" in possession-getting ready for "Sakela Sili Ubhauli Festival'

Khunda

Prior to the coming of firearms i.e. up to the early 18th century, Nepali Gurkha warriors would be armed with *khunda* (also referred as 'Konra' or 'Khura' in earlier days). *Khundas* are large, heavy and curved. Back of its edge at the hilt part, it has a width of 1½ inch. Whereas its width can expand up to 6 inches at point near which it bears an eight-petal lotus flower incised in steel. Circular disks protect the hilt, and its wooden sheath is covered in leather. Usually, it has 2 ft. 5 inches length. Its handle is made of a single iron piece. Its sharpened blade has inside edge.

Since it does not seem similar to the one available in India and other neighboring areas, it is taken as true native weapon of Nepal. Other weapons such as *talwar* and *katar* are found both in India and Nepal.

Khunda is carried across back, on the left side with the sling draped across chest or over left shoulder. These days, it is used during Dashain festival for sacrificing buffalos and other animals especially on the ninth day of the festival that is known as *navaratri* or *kalaratri*.

Ramdao

Ramdao has a broad, heavy and massive blade that terminates in a kind of axe-like projection. It bears scroll ornaments, inscriptions and the Buddhist emblem of human eye, all incised in the steel. It has colored, checkered ebony handle that is often brass-mounted. It is also used chiefly for sacrificing animals. It has 2 ft. 7 inches length, and its breadth varies from 3 to 5 inches.

Shield

In earlier days, the '*Naag-Bani Dhal*' or the Snake-Hooded Shield was used as a means of shield. It would be made from black rhinoceros hide enriched with copper gilt mounts. Observing its detail art works, one finds that its center is occupied by interlaced triangles enclosing an eight-petalled lotus flower, and surrounded by eight variously chased bosses with pendant spangles of gilt-copper. The pentagon or three triangles which one comes across on it is analogous to the hexagon engraved on the seal of Solomon. It appears that such symbols could have been used as lucky signs. It has 2 ft. and 2½ inches diameter.

Kharga

A *kharga* is a long, straight double-edged blade sword strengthened with raised gilt steel supports. It has 3 ft. 3 inches length. Its width varies from 1½ to 2 inches. In earlier days, *kharga* was taken as the main symbol of goddess Kali. This was the reason why it became the sign of the kings of Nepal too.

Katari or Dagger

A *katari* is a triangular shaped, short-pointed, two-edged knife. Hilt shaped like the English alphabet H, it is usually decorated with precious ornaments. It is usually placed concealed inside the waistcoat as a secret weapon. Its length varies from 12 to 14 ½ inches.

Manufacturing Kikes

Villages populated by Kirantis usually have their metal-smiths, locally called as Kamis, though a derogatory term now. People of this profession manufacture kikes, talwar, swords, khundas, axes and other metal implements for local populace. The steel used for manufacturing such items in the past was either locally produced or traded from Tibet. By looking at the manual work and process the humble blacksmith follow to manufacture kikes, one can say that it is a coarser affair. They have to sharpen often heavy and clumsy blade, and make a heavy wooden scabbard.

The best steel required for manufacturing kikes comes from motor vehicle's spring, railway carriage springs, fishplate and sections of railway tracks. They are supposed to be ideal materials for kike's blade. I have observed and seen Nepali blacksmiths using such kinds of materials throughout Nepal. What I can say about this is that the finest continental steel is the best one for making kike. With it, one can make kike of highest quality, and stunningly fluted and damascened.

All kikes are crafted and trued-up by blacksmiths, the artist makers. Of course, these days one can get kikes made in some self-contained mini factories. Such kikes look ordinary, small, hand-powered rotary blower, and can be recognized as forged copy. But people favor to buy the one made by local blacksmiths. One should recall here the ambience such artists folks work in their *aaran* or working station. They are seen sitting next to the leather-bag they pump by their left hand while they hold steel to fire with support of prong. The grinding wheels powered by push-pull leather straps look magical. All these items are homemade and fabricated right there. The wheels are coated with mixture of resinous gum and sand. They mix, blend and melt it like carborundum pizza.

The kike begins its journey from a piece of high-grade steel. A chunk of railway track being the materials of choice although the truck springs serves nearly as well. A length of such steel is sliced from railway tracks, and repeatedly heated and hammered alongside the charcoal-fuelled forge. Finally, the blade is annealed, and fine-tempered with water poured from kettle or teapot. The forges are little more than hole on ground and serve as the store. The incandescent billet of steel passes from furnace to anvil and back. The black smith's technical skills are reflected in the tempering process. They closely guard the secret of tempering. After the blade comes in its proper and original form, then they set to prepare or make pommel from either the same or other material such as hardwood or horns.

There is a belief that kike in ancient time would be religiously assembled at Kiranteshwor, the famous place of Kirantis during the time king Yalambar ruled. Kikes assembled in such place and manner were supposed to bring victory to the warriors. Such was the belief.

Plate B

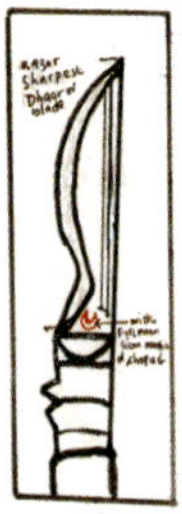

Plate C

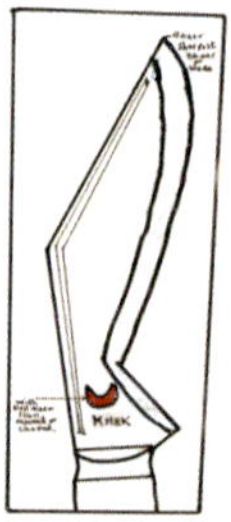

Plate D

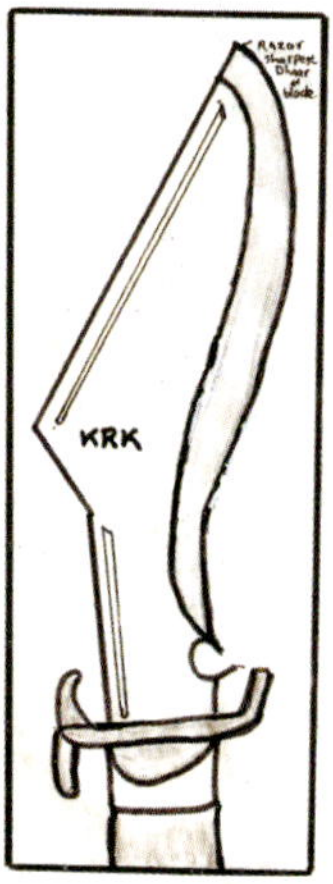

Plate E

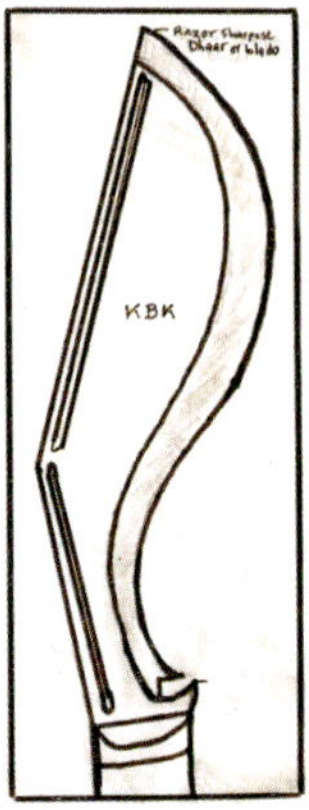

Plate G

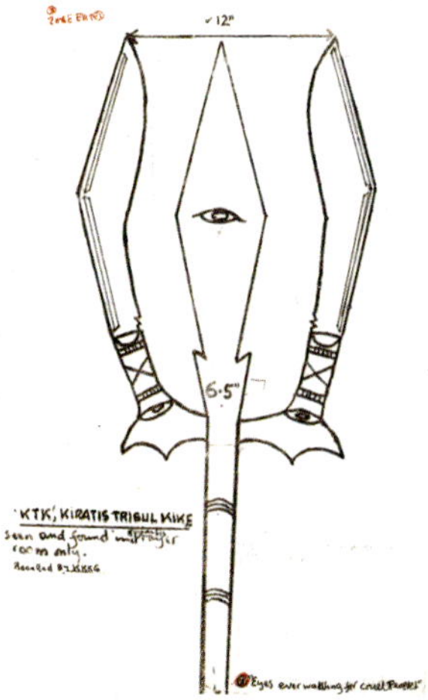

Figure No. 7: Designs and styles of Kikes used by Kirantis' ancestors (See Plates A, B, C, D, E, F and G) (Note: Plate F not available; available upon request) (Source: KKKG)

Plate 'A'

This ancient kike is believed to have been used by the legendary king Yalambar Hang's warriors, the Thulung Rais of Solukhumbu, east Nepal. Its length varies from 10 to 14 inches. Since its blade has 5 to 6.5 mm width, it can weight 2 kg and more.

Plate 'B'

It is also an ancient kike used by Kirantis from the middle Kirant region. It has same length and weight as the one from the plate 'A'.

Plate 'C'

KHHK (Kirantis' Haka Haki) style of kike has a half moon's sign on it. These days, such kike is found only in the prayer room of some devoted Kirantis. It has same length and weight as the one from the plate 'A'.

Plate 'D'

Ancient Kiranti Rais would use 'KRK' (Kiranti's Raksha Kike or Kiranti's Protective Kike) for ritual occasions. This type of kike has same length and weight as the one from the plate 'A'. It is placed in the '*Manghim*' or prayer room of any devout Kiranti family.

Plate 'E'

'KBK' (Kiranti's Bhure Kike) is especially used for cutting, thrusting and slashing. Its length varies from 12 to 14 inches. Its blade has 7 mm breadth, and it usually weights 2 kg. It is also placed in the prayer room of devout Kiranti family.

Plate 'F'

'KSTK' (Kirant's Sarpa Tauke Kike or the Snake Headed Kike of Kirantis) too is placed in the prayer room of devoted Kiranti families. Kiranti warriors in the past would use it for hooking purposes.

Plate 'G'

'KTK' (Kiranti's Trisul Kike) or the Trident Shaped Kike of Kirantis is also enshrined in their sacred prayer room.

With this, I would like to request my reader to regard khukuri or kike in terms of myths, rituals, day-to-day life activities and historical and political roles it is played. Khukuri or kike is more than a weapon of Nepalis.

Chapter III

Historical and Symbolical Values of Khukuris

History of Arsenal in Nepal

At Lagan, the midst of the old city of Kathmandu, sits an old palace that was the ancestral home of the famous Thapa family. The most famous member of this family was Bhimsen Thapa, who is regarded as Nepal's first prime minister. He remained in office for some 30 years. During his times, he reorganized the Nepal Army on modern lines. He made cantonments, magazines and barracks and drilled the Army into hard military training. To make the Army men feel proud of their ranks, he made badges and headdresses and uniforms. In the late 1830s, Bhimsen retired to his country estate. However, in 1839, his enemies lured him back to Kathmandu and imprisoned him. The charges of high treason were laid against him, and to break his spirit, he was informed that his wife for many years was to be dragged naked through the city. The night before this outrage was to occur; Bhimsen hanged himself in his cell. After hearing the news, his wife threw herself from a high window of their palace with a rope around her neck. Even today rope hangs from the very spot in homage to this great statesman. In 1845, Bhimsen Thapa's nephew became prime minister and initiated a full inquiry, which completely exonerated his late uncle. A great bloodbath followed, eliminating all the original Thapa enemies. Lagan Silekhana was never got reoccupied by this family but became the royal armory. From this particular armory, the Atlanta Cutlery Corp and the International Military Antiques of New Jersey purchased many rare, original weapons. Many such items had not been available for decades. This was truly their great opportunity to add to their British Victorian Weapons in their weaponry history, which also promoted the British Colonial and the US Civil War Collections.

There had been a crude kind of arsenal at work in the Valley as early as 1770! During Nepal's first Prime Minister Bhimsen Thapa's period from 1806 - 1837, magazine was established in Pyuthan district of Rapti zone with 500 armors of Khalifa clan. It was decided so because the best quality irons and metals were found only in Pyuthan, which then was best for making rifle's barrels and for kike's blades. Provisionally, these weapons were made there and finally sent to Tukucha in Kathmandu for final boring, assembling, fittings and disposals for the Gurkhali Armies.

The then Prime Minister Deva Shamsher had taken measures to manufacture under the guidance of Gehendra Shamsher the 8000 Martin Henry - type rifles and 6 batteries of 7- pounder guns when the British wanted to put more stringent restriction on arms supply to Nepal. These progressive ideas seemed harmful to the interest of his successors, so, on the plea that he neglected the state affairs and indulged himself in pleasure and luxury leaving aside the works in the hands of the 'Mukhtiyar' or Army General, he was removed from the premiership and sent into exile. He was banished to Dhankuta, and later on he escaped to India. While Chandra Shamsher was in Delhi in 1908, he tried to plot against him, but could not succeed. He died in the month of February 1914.

Nepalese Arms

An Illustrated Handbook of Indian Arms and Those of Nepal, Burma, Thailand and Malaysia, originally published in London in 1880 catalogues several traditional weapons of Nepal. I would like to cite a whole page from the book to make readers realise the range of varieties of Nepali traditional weapons and subtypes of khukuri weapons in the then

times in Nepal. The serial number indicates the weapon listed in the catalogue:

298.	Quiver and Arrow; the quiver of leather ornamented with hemi-spherical metal bosses, and appliqué embroidery of colored leather. Length of Tir (arrow) 2 ft. 6 in.
299.	Quiver; similar to the proceeding specimen
300.	Quiver; similar to S. Nos. 298, 299, together with a leather case with waist strap for carrying the 'tir' bow.
304.	(Not on diagram) Pallet bow, 'Guleli' made from bamboo, the strings made of thin strips of the same. For hurling pellets of clay 'matengro' or stone. Still used in 'Majh' and 'Pallo' Kirants of Nepal. Bows length 3 ft. 11 in to 5 ft. 11 in.
314.	Khukuri bright blade, the hilt damascened in gold with diaper of 8-petalled conventional flowers length 19 in; width 1 ½ in.
315.	Khukuri bright blade, hilt of horn ornamented with flower carved in low relief length 23 in; width 2 in.
318.	*Khukuri bright blade, the back edge perforated at intervals, hilt of dark wood, sheath of black leather, ornamented with quill embroidery, length 18 in. Presented by the late Sir Janga Bahadur.
319.	*Khukuri burnished steel blade, hilt mounted with silver and bound with plaited silver wire; sheath enriched with massive silver mounts, chased. Length 19 in; width 1 to 2 in.
322.	'Khunda'; large, heavy, curved and grooved blade, expanding from 1 ½ in; in width at the hilt to 6 in at the point, near which it bears an eight-petalled lotus flower incised in the steel. Hilt protected by circular disks; wooden sheath, covered with leather. Length 2 ft. 5 in. Its use and carrying procedure is explained in the 'Development' section.
323.	Khunda: Similar in form to the proceeding specimens; the blade is incised with a circular conventional flower, and a small Buddhist cross. Length 2 ft 6 in; width of blade 1 ½ in to 6 in.
332.	Sword (tarawar): Watered steel blade, very slightly curved, pommel of ivory. Length 2 ft. 9 in; width 1 ¼ in to 1 ½ in.
333.	Sword-tarawar, long, straight, double-edged blade, strengthened with raised gilt steel supports. Presented by Sir Junga Bahadur. Ancient Nepalese. Length 3 ft. 3 in; width 1 ¼ in to 2 in.
335.	Daggar-Katar or Katari, slightly curved blade, grooved on both sides, the grooves once filled with seed pearls; silver-plated with hilt, embossed with floral decoration. Length 15 ½ in; width 1 ½ in.
336.	Dagger; small curved blade, agate hilt Length 10 in.

338. Dagger, "Chilanum", flamboyant blade of Damascus steel deeply channeled, the hilt covered with simple geometrical ornaments, embossed and silver-gilt. Length 15 in; width 2 in.
339. Dagger, curved blade, thickened at the point, and ribbed, steel hilt with knuckle guard. Length 14 in.
340. Daggar, "Katar", forked blade, embossed, hilt shaped like the letter 'H'. Length 14 ½ in.
344. Daggar, "Jamdhar Katari" plain steel blade, the hilt similar in shape to that of the preceding example, bears traces of gilding. Length 13 ½ in.
345. Dagger, "Jamdhar Katari" Length 14 ½ in.
348. Knife, abruptly curved, grooved blade, silver hilt. Length 11 in.
349. Elephant Goad, "Ankus" hook, embossed steel hilt.
350, 351. Sword or Axes, "Ram dao", broad, heavy, incurved blades, ornamented along the back with scroll patterns, and each bearing a representation of the human eye incised in the steel, and colored, chequered ebony pommels with brass mounts. Length 2 ft 3 ½ in and 2 ft 5 in, Length of blades, 19 and 20 in width of blade, 2 ½ in to 4 in.
352. Sacrificial Axe; "Ram dao" or "Kharga" broad heavy, massive blade, terminating in a king of axe-like projection, and bearing scroll ornaments, inscriptions, and the Buddhist emblem of the human eye incised in the steel, and colored; chequered ebony pommel, brass mounted. Used chiefly in the sacrifice of goats and buffaloes. Length 2 ft 7 in; Length of blade 23 in width 3 in to 5 in.
355. Matchlock; heavy octagonal Damascus barrel; inlaid with silver, attached to the stock by broad bands of embossed silver plate; stock of ebony, inlaid with brass ornaments. Length 5 ft 11 in.
364. Belt; "peti", composed of four circular, slightly convex discs or plates of steel ["char aina" or "the four mirror"], connected by leather straps. Worn round the waist, outside of coat of chain mail. Diameter of discs 7 in. Presented by Sir Junga Bahadur.
365. Shield: "Nag phani dhal" (Snake-hood shield) circular, of black rhinoceros hide, enriched with copper gilt mounts enclosing an eight-petalled lotus flower, and surrounded by eight variously-chased bosses, with pendant spangles of gilt-copper. A triple border of conventional floriated scrollwork, spangles bosses, and low-relief arabesques of birds surrounds the shield. The whole is enclosed by a coil of snakes, which meet at the top of the shield, with three uplifted heads. Diameter 2 ft 2 ½ in.

This information cited texts above makes one realise Nepal's heritage of arms and weapons. Such a heritage can be said to have paved the way for the warmanship.

Industries and Markets

There are as many as 45 kike industries registered in the Government office in 14 zones and 77 districts in Nepal. Most of these industries are in the eastern and some in the western part of the country. Even retired QGOs and some senior and junior NCOs run their own mini and top kike industries in Sunsari district. They all have their skillful and professional Bishwakarmas or blacksmiths from Bhojpur, Dhankuta and Chainpur districts to run their industries. Well-known kike industries function in Pyuthan, Salyan, and Tulsipur of Dang district (Rapti zone) as well. Similarly, there are most renowned and reputed kike industries in Bhojpur. Some are also located at Dharan, Sunsari district. They are famous for making original kikes, since it is managed and operated by most retired QGOs, WOs, and senior and junior NCOs.

In the mid 90s, the Maoist's 'People's War' began. This situation affected most kikes industries throughout Nepal, so the government put restrictions in marketing, selling and purchasing of kikes openly especially in remote as well as in urban and rural areas of Nepal. Selling and buying kikes was illegalized for some time. Most kikes industries were closed down, and their finest handcrafted rare and traditional kikes had to be stored for some years.

The government has not given much attention for marketing khukuris even to this date. Many kike-industrialists have approached government for support. Some topmost industrialists have spent tens of millions of Nepalese Rupees in search of market around the globe. There are potential buyers of khukuri globally. In earlier days, some top kike sellers claimed that fifteen to twenty thousand originally handcrafted kikes would be exported abroad monthly. This proves that kike industry alone in Nepal means a good income source both for nation and industry itself. Due to the supreme characteristic of this weapon, it is exported to many countries viz; USA, UK, France, Spain, Germany, Greece, South Africa, Korea, Japan, Brunei, Singapore and so forth (i.e. before late 1990s).

Imitations

It seems as if this world is running after mass productions that imitate or copycat. But original objects and ideas remain original or natural. Apart from the working and the service issued kikes for the Gurkha soldiers, there are all sorts of imitations. Such replicas are cheap, and are found being decorated with elaborate scabbards. They have strange pommels and punched design and some slogan on blade scripted. The original and fine craftsmanship definitely deserves a higher price. However, my advice is 'BE AWARE OF IMITATION AND COPYCATS' when buying one for your own use, remember the methods of testing the blades, and *paain* used on metals, we discussed in the last chapter.

Abroad Made Khukuris

During the World War I, the Maharaja of Jodhpur, out of respect and friendship to the British Empire, raised a regiment of troops to be sent to France. Each soldier was equipped with a specially manufactured khukuri. This special khukuri had dark blue blade. Without blood gutter or kauri, it was marked with the Royal Crest of Maharajah. The pommels were made of white metal instead of usual rosewood. The scabbards were made of black hide. The blade had 13½ inches length. Its overall length was 18 inches. But it did not have original khisha, leather pouch, karda and chakmak.

During the WW I, the British officially standardized the design for Military issued khukuris in Dehradun Steel Plant in India, and some khukuris were made at Sheffield Steel Plant in England. Again, these mass-productions of machine made khukuris had no original khisha, karda, kalam and chakmak. These designs of khukuris were issued to those Gorkhas who played major role in seizing Delhi from 1828 until late 1886. It was also used mostly during the WW I, in the field and trenches of France, also during the WW II Burma Campaign and throughout the New Guinea jungles, Malaysia and Singapore. In 1951, the Wilkinson Sword Company produced about 1,400 khukuris and scabbards, which had wood grips fixed by riveting through the flat tang.

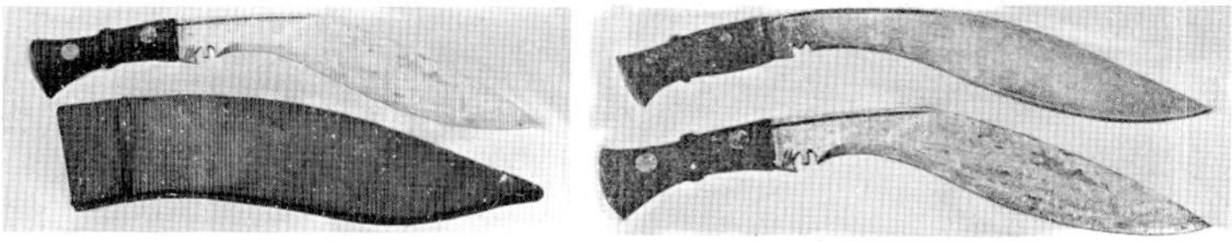

Figure No. 8: Photos of khukuri produced during the World War II
(Source: Atlanta Cutlery Corp.)

This was is the khukuri produced in a mass scale during the World War II. This machete was produced in the UK and the USA, and was carried by many of their troops. It combines the efficient cutting power of the Gurkha Khukuri with the heft of a machete made of carbon steel. Its overall length is 18 inch. Its blade has 2.75 mm × 13 mm width, and totally weights 16 Oz.

Khukuri Inspections

During the war, each group or troop commander always inspects khukuri, especially during the attacking phases of enemies' camps and their defensive positions. There are various other reasons for inspecting khukuris, but one specific reason is that it is most effective to operate it in foxholes, trenches and bunkers, where the space is so limited. In the peacetime, khukuri is always inspected during the Regimental Quarter duties by field officer of the day. During the Guard Mounting Ceremony, there is a special set of orders for unsheathing and sheathing khukuri.

Prior to 1947, Army contractors would supply khukuris, but when this was not the case, rules for acquiring khukuris got regulated for the Gurkhas. In 1906, battalions of military police were instructed to apply under an order dated 1889 to the Resident in Nepal for the necessary permit to purchase khukuris. It had to provide a roll of men who required khukuris. The British Army had to state the district in which it wanted to purchase.

According to the 'Price Vocabulary of Stores' for the Indian Army under the camp equipment, the cost of khukuri in 1914 was IC Rupees 12 each. And, no scabbard as such was provided with khukuri one bought.

In the times of War, means of supplies other than through contractors had sometimes to be resorted. The then Captain B. R. Mullaly, the historian of the 10th Gurkha Rifles writes: "The Training Depot of the 10th Gurkha Rifles at Maymyo, Burma, in 1914-16 provided drafts for the 2nd Battalion in Gallipoli and Egypt. Khukuri gave much trouble after a while inflated prices (up to 10 shilling a piece) were charged by contractors in Ghoom, Darjeeling and Gorakhpore between Lucknow and Patna, and even so far away as Hassan Abdel, and presently the supply ceased outright. The Depot having represented the matter was told to "make their own arrangement" and commenced to make their own. It was found possible to turn out eventually over a thousand well-made Khukuris and scabbards at IC Rupees 3/8 a piece.

Carriage springs would be received from the Indian Railway as raw materials to make khukuri. '*Yandeik*' and '*Simal*' trees would be brought from jungle to make handles and scabbard linings. Charcoal for the forges would be made at Kywanadauk by a standing fatigue party. To begin with, goatskins from bi-weekly market were brought, tanned and used for scabbard coverings; these were subsequently covered by leather obtained from Rangoon.

I once visited these traditional home own factories of khukuri on my first 6-month leave from Barackpore, Kolkata to Ghoom Pahar in 1957. Unfortunately, I could not meet with the late Colonel B. R. Mullaly 'Saheb' of my own regiment. He believed that "To serve with a Gurkha Soldier under the British crown was, and is, a rare privilege which nobody who has started it can ever forget".

Until 01 January 1990, all military issued khukuris were made in the contractor's factory at the British Gurkha Depot at Dharan, Paltan Bazaar. Over 3,000 khukuris were contracted annually.

A British woman, everyone called her memsahib recalls the atmosphere of Paltan bazar: "At the back of Paltan Bazar was a Khukuri factory, which seemed almost purposed-built to astound visiting Generals. It consisted of selection of ramshackle sheds where dwarfish figures that seemed to spring straight from the pen of Walt Disney constructed awesomely efficient weapons from old railway tracks and car-springs on a variety of forged and machinery that appeared to have descended from the middle ages".

Official Users and Carriers of Khukuris

1. The Nepal Army and the Armed Police Force
2. All the British Gurkha Armies
3. All the Gurkha regiments of the Indian Army
4. The Gurkha Contingent of the Republic of Singapore Police Force
5. The Gurkhas Assam Rifles Regiments
6. The Pasokan Simpanan Gurkhas (Gurkha Reserve Unit)
7. The Jordanian's Special Forces and personal guards of King Abdullah of Jordan

Carrying Khukuris

During ancient Kiranti times, traditional method of carrying kike was to tuck it in *patuka* or waistband. One can call it a herdsmen's method. According to this method, kike is tied on waistband, and slung on left side in a sash manner. But *phaudari* or combat style is to carry it secretly hidden inside left / right side of the waistcoat.

In Nepal Army, the universal custom of carrying traditional khukuris individually began during King Prithvi Narayan Shah's rule. The soldiers used to carry their khukuris with plain bright blade, black wooden pommel and leather sheathed. Such khukuris usually had 19 inches length and 2 to 3 inches width. Most officers had their own personal kike with burnished steel blade; pommel mounted with silver and band with plaited silver wire; sheath enriched with massive silver-mount.

According to the history of 10th Gurkha Rifles, the 14th Battalion of coast soldiers was the first to carry their khukuris. Between 1813 and 1837, the Tower Musket was issued with both their traditional khukuri and khukuri-bayonet musket individually. Even today, the Rais and the Limbus carry their personal combat kikes brought from their own villages. Such kikes are already tested and adjusted in their hands prior they leave for abroad. Much later, this custom of carrying khukuris spread in time to Gurkhas serving in the Army of British Raj and in Indian Gurkha. Many other hill units such as Assam Rifles Regiments, Burma Military Police and Garhwal and Kumaon Regiments carry it. During the Burma Campaign in the WW II, those British troops who did not carry machete would carry a kike. In the 11th Gurkha Rifles of Indian Army, the JCOs (Junior Commissioned Officers) carry their kikes with metallic pommel whereas others carry kike with wooden pommel.

Khukuris as Gifts

A Kothi Muhunda or Kothi Mohora Khukuri is especially made as presentation gift to the distinguished persons. It is also offered to persons of all status held in special esteem by honor and even to groups who are to be honored.

Its pommel is made from ivory, silver or rhino's horn. Its sheath is covered with green or blue or red velvet. Its top is artistically handcrafted with beautiful kothis of King's crown, peacock head, lions, elephants or even of a Gurkha soldier's figure. The top man's khukuri incorporates exquisite etching and in addition to silver or gold scabbard decorated with precious gems. To receive a Kothi Muhunda Khukuri as a gift is to receive a great pride and honor. It is equal to receiving a national icon, national weapon, national souvenir or national symbol, of an exceptional recognition. History shows Kings, Queens, Generals, and Ministers display khukuris to demonstrate power, money and pride.

Khukuri as Awards

Every year, the HRH Prince of Wales awards a khukuri to an individual in the Brigade of Gurkhas. Such an awardee is deemed to have made significant contribution to enhance Brigade's reputation. The award is open to all ranks below the Warrant Officer Class I.

On the historical day, 31 December 1962, the Pipes and Drums of 1st Battalion 10th PMO Gurkha Rifles bid a farewell retreat on the BN Drill Square at Majedee Barracks, Johore Bahru in Malaysia to say "good bye" to barracks which had remained regimental home for 10 years. On this

Figure No. 9: Photo of Sgt. Toya Bahadur Gurung with the Prince of Wales's Khukuri

special occasion, this Regiment presented a most beautiful Kothi Muhunda Khukuri with ivory pommel and artistically silver dotted 'kothis' on its scabbard to the Sultan of Johore. The Sultan, in return, presented the Battalion a beautiful kris, a Malaysian khukuri with ivory pommel.

As I remember, 10th GR's, the then Pipe Major Kishan Singh Thapa wore the most beautiful silver ornamented Kothi Mohora Kike with ivory pommel at this final farewell retreat parade.

On 28 February 1956, at a ceremonial parade in Johore Bahru, South Malaysia, the Commissioner of the Police Federation of Malaysia received the salute, and presented a handsome salver to the Battalion and accepted a Kothi Muhunda Khukuri as the gift on behalf of the Police.

In February 1956, a most beautiful original, rare and historical Samurai sword once captured by the $1/10^{th}$ Gurkha Rifles in Burma War was presented to the late Major (GCO) Purne Rai, SB, DSO, OBE, OBI, MBE, MC and Star of Nepal. This historical sword was presented to him through the hand of the then Lt. Colonel DDM McGredy DSO, OBE. He is the first and to date, the only Gurkha ever to win the Distinguished Service Order in the Britain's Brigade of Gurkhas. "The

Khukuri had always represented the spirit of a Gurkha warrior who was always willing and ready to fight for the Great Britain, anywhere, at any time, against any opponent and under any condition", opines Purne Rai. This sword is with this author at the moment.

Figure No. 10: Photo of the Sword that was presented to Major Purne Rai

A most beautiful Kothi Muhunda Khukuri with ivory pommel and highest standard of silver work on the scabbard with the badges of all Gurkha Units of the Brigade affixed to them was presented to the Queen's

Own Rifles of Canada on behalf of the Brigade of Gurkhas by Major RW Willis 7th GR as a Brigade Welfare Officer in Nepal in 1983.

Prince Harry joined 'one of the most fearsome units in the British Army'- the Gurkhas. The 22 years old Prince then was the Second Lieutenant. He was reported to have been presented with an honorary Khukuri, the curved national knife of Gurkhas. "Harry will be told that when the Khukuri leaves its sheath it can never be returned without enemy's blood being drawn" mentions *The Sunday Mirror.*

Religious Significance of Modern Khukuri

The British colonial administrators and military staff during the World War I and II considered the ritual practices associated with kike or khukuris as superstitious, demonic and pagan. Many senior Kiranti officers would share this experience with us. The British Military Officers regarded such rites as sign of the culture of backward and simple people of the hills. Many shamanistic rituals with ancient khukuri's systems were banned from Gurkha Regimental protocols. But shamanism for Nepalis is something different. Shamans and priests, healers and astrologers are part and parcels of Nepali cultures. This tradition is thought to date back long, and is still practiced in Nepal and her neighboring countries.

Khukuri still is a religiously significant weapon as it is worshipped during the Dashain, the most important Hindu festival. It is worshiped at times before any sacrifice is made. Dashain is given importance in the Indian Gurkha and also the British Gurkha regiments. Since weapons are taken as symbols of power in Nepal as well as in the Indian and the British Gurkha regiments, all heavy and small arms are made sacred, and many male goats and buffaloes are sacrificed especially on Mahanawami Day (ninth day of Dashain). In Nepal, 54 male goats and 54 male buffalos are sacrificed at the Hanuman Dhoka, Dashain Ghar and Kot as well as 108 male goats and 108 buffalos are sacrificed at the Taleju Bhawani on Kalaratri night or the darkest night, i.e. Ninth day of Dashain.

The British Gurkha Army and the Nepalese Army follow same rites and rituals during the animal sacrifice given on festival like Dashain.

Their proven expert ceremonially sacrifices a fine male buffalo and goat in each regiment. A gigantic khukuri of over 30 inches length, weighing over 5 kilograms is used as the scapegoat's head must be clearly served with one clean blow. Sometimes a *khunda* is also used. So much depends on the sacrificer. When this is achieved, with 100 percent success, it is deeply believed that the Goddess Durga Bhawani keeps on showering blessings on regiment throughout the year. With this ritual, regiment's moral power reaches to the highest point.

Finally, it is also a custom to honor expert sacrificer. Usually, commanding officer and Gurkha Major offer some cash reward to him. This honor is much valued among Gurkhas. This practice has ceased in the British Gurkha Army since 1973 but not in the Pokhara Camp where a male buffalo and a male goat are ceremonially sacrificed in the name of present Royal Gurkha Rifles (RGR).

My personal experience is that Dashain has always remained a favorite festival among Gurkhas and their officers. My Maj. Gen. C.H. Boucher, CB, CBE, DSO would wish Dashain greetings in this manner:

> "I wish all "Gurkha Officers" and "Gurkha Other Ranks" that this Dashera be joyful, prosperous and good from the deep of my heart.
> Unfortunately, due to the full job of dominating 'Communist Bandits' you did not receive full 10 days leave. However, all of you may get up to 7 days leave. Although you are far from your homeland, I hope you will celebrate the festival happily".

For Malaysia based Gurkhas, 2,000 gallons of duty free rum used to be available tax free on the occasion of the festival of Dashain.

Testimonials issued by the Prime Minister and Supreme Commander in Chief of Nepal in 1947 states that the religious significance of khukuri must not be forgotten. In 1948, Maharaja Padma Shamsher JBR, the Prime minister and Supreme Commander of Nepal wrote: "The Khukuri is the national as well as the religious weapon of the Gurkhas. It is incumbent on a Gurkha to carry it while awake and to place it under the pillow when retiring. As a religious weapon, it is worshipped during

Dashain (the most important Hindu festival) and at other times whenever any sacrifice is to be made".

Marking Khukuri

One finds carved figures of the sun, the half and the full moons, the holy Devanagari alphabet '*shree*' and some unknown deity laid in gold and brass on kike's blade. Different parts of kikes are found to have the stamps of the WW I & II. It is not possible to interpret with confidence, who could have stamped such markings. Some refer to unidentified contractors' marking; others refer to regimental issue and date of issue etc. It is also probable that marks record a soldier's personal or army number. One example found thus: A kike once preserved by 1/10th PMO Gurkha Rifles may be one of this. This kike was picked up in Suvla Bay in August 1915 by a Royal Navy Sailor from a dead Gurkha of 2/10th Gurkha Rifles, and was returned to the Regiment by the sailor in 1965. It is of interest that the kike is simply made without a pommel plate and bears the number 725 on the back of the blade. This almost certainly refers to the 725 Rifleman, Ran Bahadur Rai who was killed on 20th August 1915, when the Battalion was in the Suvla Bay area. I personally have cleaned and taken care of this particular kike that was displayed in the Regimental Guardroom of my own Regiment in Malacca, Malaysia in 1963. Today, this reminds me of "our kike, our pride."

John, the king of medieval England was unable to write, so he did not sign the Magnacarta (the great charter guaranteeing liberty) in 1215. He stamped it. In the same manner our Kiranti Kings should have done the same in earlier, ancient times so that today we could claim our rights. Even today, for various uncertain reasons, the issued kikes to the soldiers of all armies do not show marks or stamps. It is my personal view for future generations to have a specific mark for kikes made in Nepal from 21 century onwards.

Among khukuris locked and stored in old Nepali palace, some are over 100 years old. Each khukuri in such collection is hand-made. Its steel was melted, forged, shaped and hardened in Nepal by many skilled Nepali blacksmiths. They are the original blades. These form of khukuris have been in the service of the Nepalese Army, Indian Gurkha Rifles and British Gurkha Army. Some blades designed in earlier times have names of gods carved in the Devanagari script. However, in the khukuris manufactured during the World War II, one can find of manufacture years such as 1944, 1944 and 1945 being scripted in the blade. It is believed that the Atlanta Cutlery Corp, and the International Military Antiques of New Jersey have purchased many rare and original weapons from the Royal Nepalese Army during the reign of king Gyanendra Bir Bikram Shah. I guess the collection is sure to have some original khukuris used by our Gorkhali warriors to defeat their enemies since the 1814 war and also during the WW I & II. The collection that is supposed to have bought by the USA based organisations mentioned above is sure to have many khukuris that are over 100 years old. There can also be khukuris made much earlier though it's difficult to determine the exact date because of doubtful markings on blade itself. There were khukuris made for war, especially for close-quarter combat.

***Antique* Nepalese Kukris**

Figure No. 11: Picture of antique Khukuris with markings
(Souce: Atlanta Cutlery Corp.)

The following are the khukuris used in the Nepalese Military as well as the British Army. Many blades have marks scripted in the Devanagari script.

(A): It is a traditional khukuri with its unique *sirupate* or long leaf design. Shiru is a grass that has long slender leaf. This khukuri has approximately 19 inches length. It has 2½ inches width, and can have 8/3 inches thickness. As said earlier, this is the khukuri Kirantis love to carry.

(B): This is famous traditional Bhojpure khukuri. It has approximately 17 inches length. It has 2½ inches width, and can have 8/3 inches thickness. Mostly the Gurkhalis use it.

(C) & (D): This is the famous WW II vintage khukuri. Later, slightly smaller version of traditional khukuri was issued. Most of the WW II khukuris are dated with the year of manufacture: 1943, 1944 and 1945.

(C): This khukuri has approximately 17½ inches length. It has 3/8 inches width and 9/32 inches thickness.

(D): The Gurkhas used this khukuri to siege the Delhi Sultan in 1857. It has aproximatley 17 inches length, 1/8 inches width, and 1/8 inches thickness.

(E): The WW I issued khukuri has approximately 17 ¼ inches length, 3/8 inches width and 13/32 inches thickness.

When did the marking of khukuri begin we can not exactly say. Some refer to unidentified contractors to the Indian Army; others refer to regimental issues and dates of issue. It is also probable that marks record a soldier's personal number.

Plate 1

Figure No. 12: Plate 1 (Source: National Army Museum London (DFH))

No 1

The marks F W on the hilt here may refer to the Fort William Depot although the letter W often refers to 'Workshops'. The 1.15 refers to January 1915 when it was inspected. Similarly, the year 1919 has also been scripted (National Army Museum 8307-102).

No 2

The marks COI and arrow over G may refer to contractors. I.G may refer to Inspectorate Gorakhpore, which was an important Gurkha recruiting centre in the past. The presence of broad arrow over 'I' is generally constructed at this period to indicate British manufacturer for Indian service. The mark 1917 on the blade and scabbard indicates either the date of manufacture, acceptance or issue. The 'I' arrow, M over S refer to Indian Military Stores (National Army Museum 8205-19).

No. 3.

The mark DHW may again refer to a Government workshop. This is followed by the 'I' and the date 1920. These markings are similar to those one published specimen dated 1919 (National Army Museum 8404-138).

No. 4.

The blade is engraved with a device of Bihar and Orissa Military Police. The scabbard is marked with date 31 August 1932. In all its features, it is far superior to the Gurkha issued weapon of the time both in quality of its forging and quality of hilt and scabbard (National Army Museum 7406-21). One such 1915 dated khukuri, found in a private collection, has frog markings attributable to the 2nd BN, 39th Garhwal Rifles and regimental number of the soldier 409, when it was issued.

Plate 2

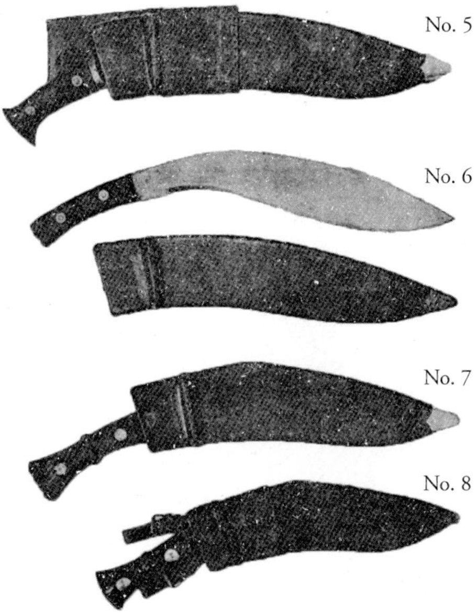

Figure No. 13A: Plate 2 (Source: National Army Museum London (DFH))

No. 5.

Dated 1945, the blade is stamped ATD 42. This may refer to the Army Traders Dehradun 1942. The scabbard bears the date 1945, the frog bears the date 2/43, indicating the date of inspection, i.e. Feb. 1943. In an illustration of a Gurkha soldier taking part in the Italian Campaign of the World War II. The scabbard of a similar khukuri is found protectively wrapped in a cloth bandage for protection or camouflage (National Army Museum 7505-2).

No. 6.

Its blade is made of slab steel without any fuller. Its edge has an acute profile without notch. Its grip looks like that of more jungle-knife. Unlike that of the usual khukuri, it is riveted with issue pattern rivets. Its scabbard has all attributes of an issue pattern. The stamps appearing on the specimen include SA / 150, which is probably an inspector's mark, and S.C.W. / 43, which probably refers to a specific workshop and the date 1943. It is known that khukuris were made in Indian Railway workshops at this period. As is usual, a brass cap was stapled to the scabbard and it was painted as khaki, a common practice then (National Army Museum 8402-68).

No. 7.

The blade stamp HW45 probably refers to a workshop and the date 1945. A further date of 1945 with COC is stamped on its scabbard (National Army Museum 8304-104).

No. 8.

It is a service-issued weapon with buckled belt loops and a frog. Its short blade with 26.2 cm can be contrasted with the normal 31 to 35 cm. Its 425 gm. weight can also be contrasted with the one with 560 to 640 gm. weight. Its grip, like that of the present service khukuri, is made of horn but it is without marks (National Army Museum 5904-340-2). It resembles another khukuri in the collection (National Army Museum 6804-8-2). The blade of which is inscribed: The Army Traders Dehradun.

Plate 3:

Figure No 13B: Plate 3 (Source: National Army Museum London (DFH))

This khukuri was issued for the British Gurkhas. It has a tang of blade riveted through pommel's cap or chhapri unlike most of the WW I & II issued ones. It is made of crude blade plate, and its *kauri* has a simple letter 'E' form.

The scabbards have laced *namja* frogs. They include *karda* and *chakmak* but not old traditional *khisha* or a leather purse. Without traditional *khisha*, the leather usually has blocked ridge to prevent scabbard slipping through *namja*.

Present service version does not carry marks, and is slightly smaller than the earlier specimens. The blade is an average 20.5 cm long (chord) in comparison with the average of 31 cm of the First and the Second world war variant.

Awards and Bravery

No war; no awards. One day, an orderly officer at barrack noticed a soldier wearing three medals of courage. Impressed, the officer went to him and asked, "Where did you get those medals?" The soldier looked at him for a moment and then replied, "From the central store, sir!"

This reminds me of a retired Indian Gurkha Army Major, a close friend of late Major (GCO) Purne Rai. One day he asked me of what bravery I had received so far. I simply explained to him not other than of my three medals – MP, GSM, and LSG, and added "Hajoor, I am expecting to receive the highest and a longest 'bravery award' from the Sultan of Perak (Malaysia) in near future, the award spells: SOMPKHPBSKJCDPMWDTB. Hearing this, he was so surprised and not knowing that this in fact is the longest Malaysian word phrase, and after a while, he repeatedly said, "great Saheb, one should be proud of you, syabash!"

Later, I had to explain him the full meaning of each letter, thus;

Syarikat Orang Orang Melayu Pegawai Kerajaan Hilir Perak Bekerja Sama Sama Kerana Jimat Cermat Dan Pinjam Meminjam Wang Dengan Tanggongan Berhad, i.e. "Company of the Malay people and officers of downstream Government of Perak work together in order to save wisely and borrow money with guarantee limited". Many words are spelt in the old Malaysian language.

Khukuri and the Victoria Cross

Instituted in 1856 in the name of the Queen of Great Britain in (1837-1901) and the Empress of India (1876-1901), the Victoria Cross was and still remains the most prestigious British award for gallantry. It can also be given to a civilian under military command.

A First Appraisal of the Victoria Cross Medal

On 26 June 1857, Queen Victoria held a review in Hyde Park at which she presented some Crimean veterans with the first awards of her newly instituted Victoria Cross. *The Times* wrote disparagingly of the medal:

'Never did we see such a dull, heavy, tasteless affair … there is a cross and a lion and a scroll worked up in the most shapeless mass that the size admits of. Valour will have to be its own reward; had the medal been on sale at one penny, hardly a dozen would be sold in a twelve-month'.

However, it is recorded elsewhere that Her Majesty liked it. And, we know what the world thinks of it today (32), writes *The Bugle and Khukuri*, the journal of the 10th PMO GR 2002.

Her Majesty Queen Elizabeth II is the patron of the Victoria Cross and the George Cross Association currently.

It is still top of the wish list for the most serious medal collectors in the British Army. Each one bears an incredible story, and is not only highly valuable but also of great historical significance.

It has been so far awarded 1,356 times and instantly recognizable with its crimson ribbon and depiction of all lion in the center of the cross itself. Remarkably, the jeweler, Hancocks of London has produced every single VC since its inception.

The VC holders are scarce. Only 13 Gurkhas have been awarded since the end of the WW II.

London specialist Dix Noonan and Webb have sold 22, the most recently when a British record £ 180,000 was paid for a WW I VC given to Corporal William Cosgrave of the Royal Munster Fusiliers in Gallipoli. But £ 500,000 was paid for one in Australia.

Medals have a huge international market with quite a few younger collectors. The US Congress passed laws to make it illegal to sell American medals and decorations, writes *The Sunday Mirror*.

Until 1912, only the British officers of the Gurkha Regiments were eligible for receiving the Victoria Cross. Gurkha officers and men had been awarded by the Indian Order of Merit for gallantry. King George V's order of 1912 made all officers and men of the Indian Army eligible for the supreme award (30), writes Mike Chappel in his book titled *The Gurkhas*.

Since 1911, Gorkhas were also considered for this prestigious British award. Only those individuals who have performed greatest heroism in extreme danger are considered for the decoration. The citations of nonstop accounts of disregard for bombs and bullets in order to rescue a comrade or gain an objective are almost unbelievable. Amongst the thousands of bravest of brave Gurkhas performing their bravery during the WW I and II, fortunately only 3 of them had their opportunities to receive it. They deserved it because they had demonstrated and proved their kikes' actions on real battlefield.

Figure No. 14: Photo of the Victoria Cross (Source: Wikepedia)

The truth of the belief that a Gurkha must draw blood every time after he unsheathes his kike (khukuri) in battle was set by these three under mentioned Gurkhas in the WW II:

1. Captain Lal Bahadur Thapa Magar 2/ 1st GR
2. Captain Netra Bahadur Thapa Magar 2/ 5th GR
3. Bhanu Bhakta Gurung 3/ 2nd GR

1. Capt. Lal Bahadur Thapa Magar

Figure No. 15: Photo of Capt. Lal Bahadur Thapa Magar, 2nd/1st GR

Capt. Lal Bahadur Thapa Magar, 2nd/1st GR VC OBI, was the first Gurkha officer to receive the VC in 1943. Citation from the London Gazette that describes about the bravery of Capt. Magar goes:

> "On the night of 5th and 6th April 1943, during the silent attack on the Rass-Ez-Zouai feature, Capt. Lal Bahadur Thapa was second in command of "D" Company. The commander of No. 16 Platoon was detached with one section to secure an isolated feature on the left of the Company's objective. Capt. Lal Bahadur Thapa took command of the remaining two sections and led them forward towards the main feature on the outer ridge, in order to break through and secure the one and only passage by which the vital commanding feature could be seized to cover the penetration of the Division into the hills. On the capture of these hills, the whole success of the Corps plan defended.
>
> First contact with the enemy was made at the foot of a pathway winding up a narrow cleft. This steep cleft was thickly studded with a series of enemy posts, the inner of which contained an anti-tank gun and the remainder medium machine guns. After passing through the narrow cleft, one emerges into a small arena with very steep sides, some 200 feet in height, and in places

sheer cliff. Into this arena and down its sides, a number of automatic weapons were trained and mortar fire directed.

Capt. Lal Bahadur Thapa and his men killed all the garrison of the outer posts by khukuri or bayonet in the first rush and the enemy then opened very heavy fire straight down the narrow enclosed pathway and steep arena sides. Capt. Lal Bahadur Thapa led his men on and fought his way up the narrow gully straight through the enemy's fire with little room to maneuver, in the face of intense and sustained machine-gun concentrations and the liberal use of grenades by the enemy.

The next machine-gun posts were dealt with Capt. Lal Bahadur Thapa personally killing two men with his khukuri and two more with his revolver. This Gurkha Officer continued to fight his way up the narrow bullet-swept approaches to the crest. He and two riflemen managed to reach the crest, where Capt. Lal Bahadur Thapa killed another two men with his khukuri, the Riflemen killed two more and the rest fled. Capt. Lal Bahadur Thapa then secured the whole feature and covered his Company's advance up the defile.

This pathway was found to be the only practicable route up the precipitous ridge and by securing it, the company was able to deploy and mop up all enemies. This objective was an essential feature covering the further advance of the Brigade and of the Division, as well as the Bridgehead over the anti-tank ditch.

There is no doubt that the capture of this objective was entirely due to this act of unsurpassed bravery of Capt. Lal Bahadur Thapa and his small party in forcing their way up the steep gully, and up the cliffs of the arena under withering fire. The outstanding leadership, gallantry and complete disregard for his own safety shown by Capt. Lal Bahadur Thapa were an example to the whole company, and the ruthless determination of this Gurkha Officer to reach his objective and kill his enemy had a decisive effect on the success of the whole operation."

Lal Bahadur Thapa Magar served as a Subedar Major in 5th/2nd GR till 1944 and got retired as an honorary Captain in 1948.

Lal Bahadur Thapa Magar VC, OBI, Nepal Tara, 2nd/1st GR expired on 20th October 1968 at the age of 62 years.

2. *Capt. Netra Bahadur Thapa Magar, 2/5th GR FF*

Citation from London Gazette

Figure No. 16: Photo of Capt. Netra Bahadur Thapa Magar 2nd/5th GR

The king has been graciously pleased to approve the posthumous award of the Victoria Cross to Netra Bahadur Thapa (28467/ 10), 5th Royal Gurkha Rifles (Frontier Force, Indian Army). The citation from the London Gazette describes the bravery of Capt. Thapa as follow:

> "Netra Bahadur Thapa was in command of the garrison of 41 men of the 2/5th Royal Gurkha Rifles (Frontier Force) which on the afternoon of 25th June 1944, took over the isolated piquet known as mortar bluff situated on the hillside commanding the base at Bishenpur in Burma. The piquet position, completely devoid of any cover, was situated some 400 yards from the next piquet from which it could be supported to some extent by 3-inch mortar fire, but was commanded by water piquet. A short distance away on high ground to the south, which had been overrun by strong enemy forces on the previous night and was still in enemy hands. Owing to its commanding position the retention of mortar bluff was vital to the safety of other positions farther down the ridge and to Bishenpur itself.
>
> The relief had been harassed by enemy snipers at close range but was completed at 1830 hours without causalities. A little more than an hour later, the enemy began to attack. For this purpose a 75 millimeter and a 37 millimeter guns were brought on up to the high ground overlooking the

position and poured shell after shell at point-blank range for ten minutes into the narrow confines of the piquet, and this was followed by a determined attack by not less than one company of Japanese. A fierce fight ensued in which Netra Bahadur Thapa's men exhorted by their leader, held their ground against heavy odds and drove the enemy back with disproportionate loses. During this time, Netra Bahadur Thapa with tireless energy and contempt for his own safety, moved from post to post encouraging his young NCOs (Non-Commissioned Officer) and riflemen of which the garrison was largely composed.

A short lull followed during which Netra Bahadur Thapa gave a clear and concise report on the telephone to his commanding officer and asked for more artillery defensive fire. Having done this, he made preparations to meet the next onslaught, which was not long in coming.

Under cover of the pitch-dark night and torrential rain, the enemy had moved round to the jungle from cover of which they launched their next attack. Still in considerable strength and as determined and ferocious as ever, the enemy poured out from the jungle across the short space of open ground to the piquet defenses under cover of small arms and 37 millimeter gun fire from flank. For a time, his men held their ground until, as ill luck would have it, both the LMG and TMG of one section jammed.

With much reduced firepower, the section was unable to hold on, and the enemy forced an entrance and overran this and another section, killing or wounding 12 out of the 16 men comprising the two sections. Having no reserve, Netra Bahadur Thapa himself went forward from his headquarters and stemmed any further advance with grenades.

The situation was, however, critical, with more than half his men were passing through casualties, ammunition low and the enemy in possession of part of his perimeter. Netra Bahadur Thapa would have been justified in withdrawing, but in his next report to his commanding officer, he stated that he intended holding on and asked for reinforcements and more ammunition.

So efficient were his plans for defense and such was the fine example of this gallant Gurkha officer that not a man moved from his trench and not a yard more ground was gained by the enemy, despite their desperate attempts.

Thus the night passed until at 0400 hours, when a section of eight men with grenades and small arms ammunition arrived. Their arrival inevitably drew fire and all the 8 were soon casualties. Undismayed, however, Netra Bahadur Thapa retrieved the ammunition and he with his platoon headquarters took the offensive armed with grenades and khukuris. Whilst so doing, he received a bullet wound in the mouth followed shortly afterwards by a grenade which killed him outright. His body was found next day, khukuri in hand and a dead Japanese with a cleft skull by his side.

True to the traditions of the service and his race, Netra Bahadur Thapa had fought against overwhelming odds for 8 hours before he was killed. His fine example of personal bravery and his high sense of duty so inspired his men that a vital position was held to the limit of human endurance.

His valour and devotion to duty will remain an epic in the history of the regiment."

His dearest wife Mrs. Naina Shara Thapa Magar received her husband's VC. She handed over the VC to the regiment. And she passed away the following year 1945.

3. Rifleman Bhanu Bhakta Gurung

Figure No. 17: Rifleman Bhanubhakta Gurung 3rd/2nd GR

The king has been graciously pleased to approve the award of the Victoria Cross to:

No. 10020 Rifleman Bhanubhakta Gurung, 2nd King Edward VII's Own Gurkha Rifles, Indian Army. The citation from the London Gazette describes the bravery of Rifleman Bhanu Bhakta Gurung goes:

"In Burma, on 5th March 1945, a company of the 2nd Gurkha Rifles attacked an enemy position known as Snowden East. On approaching the objective, one of the sections was forced to ground by very heavy light machine - gun, grenade and mortar fire, and owing to the severity of this, fire was unable to move in any direction. While thus pinned, the section came under accurate fire from a tree sniper some 75 yards to the south.

As his sniper was inflicting casualties on the section, Rifleman Bhanu Bhakta Gurung, being unable to fire from the lying position stood fully exposed to the heavy fire and calmly killed the enemy sniper with his rifle, thus saving his section from suffering further casualties.

The section then advanced again, but when within 20 yards of the objective, was again attacked by very heavy fire, Rifleman Bhanubhakta Gurung, without waiting for any orders, dashed forward alone and attacked the first enemy foxhole. Throwing two grenades, he killed the two occupants and without any hesitation rushed onto the next enemy foxhole, and killed the Japanese in it with his bayonet.

Two further enemy foxholes were still bringing fire to bear on the section, and again Rifleman Bhanu Bhakta Gurung dashed forward alone and cleared these with bayonet and grenade. During his single-handed attacks on these four enemy foxholes, Rifleman Bhanubhakta Gurung was subjected to almost continuous and point blank light machine gun fire from a bunker on the north tip of the objective. Realizing that this light machine gun would hold up not only his own platoon, which was now behind him, but also another platoon which was advancing from the west, Rifleman Bhanu Bhakta Gurung for the fifth time went forward alone in the face of heavy enemy fire to knock out this position. He doubled forward and leapt on to the roof of a bunker from where his hand grenades being finished, he flung two No. 77 smoke grenades into the bunker slit. Two Japanese rushed out of the bunker partially blinded by the smoke. Rifleman Bhanu Bhakta Gurung promptly killed them both with his khukuri. A remaining Japanese inside the bunker was still firing the light machine gun and holding up the advance of No. 4 platoon, so Rifleman Bhanu Bhakta Gurung crawled inside the bunker, killed this Japanese gunner and captured the light machine gun.

Most of the objective had now been cleared by the men behind and the enemy driven off were collecting for a counter - attack beneath the north end of the objective. Rifleman Bhanu Bhakta Gurung ordered the nearest bren gunner and two riflemen to take up positions in the captured bunker. The enemy counter-attack followed soon after, but under Rifleman Bhanu Bhakta Gurung's command, the small party inside the bunker repelled it with heavy loss to the enemy.

Rifleman Bhanu Bhakta Gurung showed outstanding bravery and complete disregard for his own safety. His courageous clearing of five enemy positions single-handed was in itself decisive in capturing the objective and his inspiring example to the rest of the company contributed to the speedy consolidation of this success."

Rifleman Bhanubhakta Gurung VC, the star of Nepal, retired in 1946. He was given a rank of honorary sergeant and no official pension. He lived in Gurkha Bazaar with his wife and daughter, and expired on Saturday, 1 March 2008, at the age of 85 years.

The Last Chunk!

But it's enough for 81 more VCs!

Jackie Brown, petite, 22 and unattached, lifted the lump of metal in the Small Arms Museum at Donnington and said: "I can pick this up quite easily. It weighs only 22 ½ lb."

Jackie, a store clerk in the Small Arms Store at the giant RAOC base, was in fact holding the last chunk of bronze from which VC medals are made.

The most valuable piece of military metal in the Army, the last bit was cut from it some years ago when the London firm of Hancocks, the court jewelers to the Queen Mother struck 12 new VCs.

The remaining 22 ½ lb. is enough to make another 81 medals, but as so few are awarded, it's likely to last a good few years yet.

While the metal waits to be reduced in size still further, it stays in the charge of Lt. Col. John Grainger, OC, the sub-depot at Donnington, which includes the RAOC's private small arms museum.

There, on a plinth supported by four upright century-old rifles, this priceless metal sits for visitors to view.

Dull yellow in colour, it resembles a bit of slightly mouldy Swiss cheese because of blowholes in the casting.

"Our last issue of metal from this cascable (the bit from the breech end of a cannon and by which it is towed) was in 1970," said Col Grainger. "Then 3 lb. was cut off and sent to the medal makers".

Originally there was 100 lb. of metal, but since the medal was introduced in 1856, more than 1350 have been awarded and this is all that remains of the original cascable.

The cannon from which the tow bar came was captured at Sebastopol in the Crimea war and is now in the Royal Artillery Museum at Woolwich.

But it's not likely that metal will be taken from that to make medals as estimates reckon there is sufficient remaining of the 22-½ lb. cascable to last 200 years!

Of course, it's the medal that makes the metal so valuable. By itself, it is practically worthless. Experts know it to be bronze, but are undecided whether the cannon is of Russian or Chinese manufacture.

The monetary value placed on a VC is high. Some are sold, but holders and families tend to keep them. It is the rarity of the VC that collectors do prize so much and the value escalates accordingly.

The last one sold at auction fetched £62,000, writes *The Soldier Magazine.*

Khukuris or kikes have carved out its space in the annals of world history, and has become the part of heroism. It has become an important image to cherish with. No doubt, it is used in ordinary activities but equally important thing is that it has played extraordinary role promoting the Gurkhas and also the peace and heroism in the world.

Chapter IV

Global Gurkhas

Gurkha, a township in mid-west Nepal, is head quarter of the district of the same name. Situated on a hill overlooking the snowy peaks of Himalayas, Gurkha has two attractive temples of Gorakhnath and Kali. Usually people coming from Gurkha district are called Gorkhalis. But presently, people of Nepal as well as those of Nepali origin residing outside Nepal such as Darjeeling and Sikkim are called Gorkhalis (Gurkhas).

Prior to the conquest by the Gorkha, Nepal was divided into many tiny states groups such as: the Baise Rajya (twenty two kingdoms) in western Nepal, the Chaubise Rajya (the twenty four kingdoms) in far-western Nepal, the Kiranti Rajya in the eastern Nepal, the Limbu Rajya also in the eastern Nepal and (four tiny kingdoms) in the Kathmandu Valley. Each of these kingdoms was again subdivided into smaller states, which were frequently engaged in quarrels among themselves. The Kirantis spread further east mainly into three main geographical bounds such as: Wallo Kirant (this part of the Kirant region) that expanded from Bhaktapur, Kathmandu eastward up to the Dudhkoshi River, (the middle part of the Kirant region) Majh Kirant that lied in between the Sunkoshi River and the Arun Khola or River, the Khambuwan 'Pockets of

Khambus and Yakha Rais' often known as "Naulakhe Kirantis", and finally Pallo Kirant (the far Kirant region) that lied in between the Arun River and the Mechi River. It was also called Limbuwan.

Unification and War

The Shah rulers of the eighteenth and nineteen centuries of Nepal had tough times as they had to fight wars within and beyond. They took over the Kathmandu Valley in the late eighteenth century. After blocking the southern and eastern parts, Prithvi Narayan Shah wished to penetrate Kathmandu and Patan for which first he had to capture Kirtipur. The first NS 880 (1760 AD) attempt made to conquer the Valley turned out to be very fatal as one of the most important warrior hero Kalu Pandey got killed. Similarly, in the second attempt made in NS 884 (1764 AD), Sur Pratap got killed. The war turned very cruel for both sides. For the Gorkhalis, it was a matter of the prestige issue. The 3rd attempt took place in NS 886 (March 12, 1766 AD). The Gurkha Army captured Kirtipur finally.

Since the war had taken lives of several heroes of the Gorkhali force, the king P.N. Shah was supposed to have got irate. Shah, the king is rumored to have ordered that the noses of the leaders of Kirtipure's people be cut off and the city be nicknamed Naskatipur (city of cut noses).

There is a controversy among historians, some say only noses of the ring leaders were cut, while others hold that all people over the age of 12 had their noses cut; still others hold that this is an exaggeration as presented by the Christian father who were expelled from the Valley at that time.

Whatever the case may be, an "inside story" told by an outsider John Parker tells this event in a quietly nondramatic way in his book titled "The Gurkhas Fighting in Faraway Places". When the original Gurkha force had laid siege to Kirtipur, their leader and founder Prithvi Narayan "sent in a message guaranteeing the safety of the town if they surrendered peacefully. When they did so, he ordered his Gurkhas to slice off the noses and lips of all males over the age of 12, in revenge for the killing of his brother Sur Pratap. A total of 865 noses were severed, and put together, they weighed more than 80 pounds (36 kilograms). Only the players of wind instruments were spared mutilation, so that they could provide welcoming music for the march past of his victorious army".

The Bagh Bhairav temple in Kirtipur was established in 219 - 246 AD. The temple is at an altitude of 1405 meters from the sea level and has 1500 square meter area. Today, the Gurkha Army's earlier weapons captured by the then Kirtipurians are still displayed at this temple. Such weapons include - swords of different shapes and sizes, kattari, daggers, khundas, helmets, prong and shields, but we see no kikes. I personally visited the battlefield areas.

I believe that war has multiple faces. And each face of war is open to multiple interpretations. The Kirtipur episode too can't be an exception. However, what one can say is that the Shah rulers weren't mere revengeful gangs. On the contrary, they knew how to maintain communal harmony with local communities. For example, people from Terhathum, Taplejung and Ilam districts in Mechi Zone, the eastern part of Nepal, and also from Sikkim in India visit Phidim, Panchthar, another place in their neighboring area on Navami, the ninth day of the Dashain festival to view some historic items presented by the late King Prithvi Narayan Shah to the denizens of the Limbuwan area.

Such an exhibition is organized by the Yonghang Limbus of Nagi, said the coordinator of the organizing committee, Ashok Khewang Limbu to the researcher. The items on display were presented in recognition of the bravery the Limbus, the victorious and ferocious fighters had demonstrated in a war against the Tibetan intruders. Such displayed items include: two religious maulo pillars, armaments and musical instruments. These items were offered as gifts to the Limbus residing in Hewa, Maiya, Tamor, Chhathar, Phak Phok and Yangbar.

Octogenarian Prithvi Bahadur Tawa told this researcher that the then king had distributed 14 such 'maulos' to the people of Limbuwan. Out of them only two are in Panchthar, one at Maisthan of Ilam, but the whereabouts of the remaining 11 pillars is unknown. Each 'maulo' is 20 feet long and weighs around a quintal.

The researcher came to hear that the late rulers from Kathmandu had gifted 10 cannons, 200 guns and seven-dozen swords. These gifts were later shifted to a safer place when some violence hit this area. People say the tradition of exhibiting such gift items began as early as 1863 AD.

Now two maulos are kept in the home of 70 Limbu community members turn by turn. A male water buffalo is sacrificed to the maulos on Navami, the ninth day of Dashami. The sacrificial buffalo must be severed at the neck at one blow, and then parts of its body are distributed ritually. The locals believe that sacrificing a buffalo pleases the Goddess, and she fulfills wishes of her followers.

After the ritual sacrifice is done, the 'maulo' is handed over to the next family. The handing-over ceremony takes place in front of the Durga temple at Tharpu-5, said an elderly local from Ilam district. A fair is usually organized on this day and people come from far and wide for prayer and entertainment. As a researcher, I visited Panchthar, Tharpu, Mechi Zone in east Nepal in October 2004.

Rise of the Heroes

In 1810 AD, Nepal was twice its present size and its encroachment by the British-India led to the first and only Anglo - Nepali War during 1814- 1816. During the period of unification, Gurkhali regiment had extended the frontiers of national state up to the Tista River in the east and up to the Killa Kangada in the west. During the wars with the British and likewise with Tibet, Nepali regiment was called Gurkha regiment, no matter from where they fought. It was only the Gurkhali warriors who rolled the heads of enemies with kike fighting resounding the slogan 'Ayo Gurkhali'.

Gurkhali warriors killed soldier of the 'British – India' regiment in the war of Killa Kangada (fort) or Nalapani in 1814 AD. This was the time they showed praiseworthy courage and bravery. On the one hand, Gurkhalis were fighting with the British in the war fort, Nepali royal conspirators, on the other hand, limited the borders entering into agreement with the British Company Government. Nepali rulers of that period had handed over half of the areas without fight to the British

Company Government, which was truly earned by the true Gurkhalis of the bygone days.

History of Nepal can't be realised without evoking the spirit of the great warriors who had set roles for heroism. In any role of honor of warriors in Nepal, the first name would doubtless be that of Balabhadra Kunwar. The fort of Kalinga, or, as it is better known Nalapani was a crude stronghold, hastily erected atop a 500-feet high hill. Its outer defenses consisted of 12 feet high double palisades of rough-hewn logs filled with stones. During the battle from the end of October and throughout November 1814, there were 600 men and women as well as a few children under the command of Captain Balabhadra Kunwar. The British had 4,000 men (3,700 Hindustani troops, 100 English troops of Irish Dragon, and others). They also had 20 cannons under the command of Major General Sir Robert R. Gillespie.

A letter asking for surrender was received at the fort one evening in October 1814. Balabhadra tore up the letter with tip of his khukuri's blade and threw it in anger. The very next day, the British responded by having some cannons carried up the hill on elephant's back. But after firing some shells, they concluded that a frontal attack was not possible. Then, General Gillespie planned an attack from different directions by four columns comprising a total of 2,730 men. Almost half of the British troops reached the walls of the fort on October 31, but they were repulsed by heavy Gurkhali's fire. The British then dragged two six-pounders up to within 30 yards of the gateway to the fort. At this stage, General Gillespie himself got shot and died in Lt. Frederick's arms.

That day, the guns within the fort played havoc among the British troops. Every single soldier in the front rank was either killed or wounded. Those few who reached the entrance were repulsed with bows and arrows, spears, swords, stones, *ghuentro* (ghuentro is a belt made from leather used for hurling stones the size of cricket balls) and doubtless received the multi-functional taste of khukuris and khundas as soft targets. Some lucky ones missed and escaped from the paths of those weapons. "The net result of this action was a General and four officers killed, besides 13 other officers wounded, 27 NCOs and men killed and 213 wounded".

On November 24, Colonel Mawby began shelling the fort with a battery of 1810 pounders from a range of about 300 yards. Three days later, a breach in the defenses was made and another attack was launched but the British troops withdrew soon, with four British Officers dead and 636 either killed, wounded or missing.

Inside the fort, there were some women and children in support of their loyal leader Balabhadra and his troops. They all had carried their *khurumbi, khurmi, hansiya* and even *kachiya* for their self-defense against their enemies. They used all kinds of domesticate weapons and food survival items edible roots, plants etc. to feed the troops. Unfortunately, no one ever thought of recording their names like those of John of Arc, Jhansiki Rani and so forth who gave their lives for their motherland.

With the supply of water cut off from the water sprout and a heavy bombardment, on the night of 30 November, the survivors numbering 70 around led by Balabhadra fought their way through the British lines with khukuris only on their hands and escaped.

John Ship wrote about the siege, "I never saw more steadiness or bravery exhibited in my life. Run they would not, and of death they seemed to have no fear, though their comrades were falling thick around them."

In a war memorial at Nalapani, the British acknowledged Balabhadra as "Our gallant adversary"- a fitting tribute indeed from an adversary.

During this war, William Fraser was appointed as a political agent and his brother James Baillie Fraser joined the party as an artist and writer. During the operations, he made drawings of Garhwal hills, which were later engraved in England by Robert Havel and published in 1820 with a text volume as "views in the Himalaya mountains."

Fraser described the figure who is third from right in this picture as Soabheer Sing, a Ghoorkha Sou Bahadur who commanded a company of second Ghoorkhas at the seize of "Jytock". The man on the left of this group was identified as 'Himmut, a Khatree village Konoor, country Kashree', west Nepal.

Figure No. 18: Photo of Soabheer Sing and others
(Source: Details obtained from the British Library's post card by the author)

Many regiments of the Royal Nepalese Army have lineages dating much earlier than the formation of the British-India Gurkhas. Amongst their battle honour, they count the wars against the British with pride. Conversely, many also show honors for service alongside the British at the time of the Indian mutiny and during the Wars.

The senior unit in Nepal Army is the Shrinath Battalion, and it was raised in 1763. The Purano Gorakh Battalion was also raised in 1763 as the First Gurkha Company. Like the Shrinath, it has an impressive record of service, first against the British and later alongside them. The Kali Bahadur Battalion was raised in 1831, and was the first Nepalese unit to fight overseas in Burma in the World War II. Other units from Nepal include the Sher, the Mahindra Dal, the Shamsher Dal, the First Rifles, the Devi Dutt and the Gorakh Bahadur battalions.

NEPALESE SOLDIERS, 1813-1837

Figure No .19: Photo of a Khukuri bayonet used by Nepalese soldiers between 1813-1837 (Source: Perceval Landon "Nepal" and The Kukri, 1994 (DFH))

Today, kikes appear in the badges of all Royal Nepal Armies, Armed Police Forces, Parachute and Artillery Regiment with over 44 ingenious varieties of arrangements.

Figure No. 20: Present day Royal Gurkha Rifle Badge that incorporates "Khukuri"

These handy knives came to use intensively during the muzzle-loading era. Perfectly sized to cut away the excess patch material before loading, these hand-forged antiques have genuine stag handles and were carried in khukuri's leather pouches by the Royal Guard of Bhimsen Thapa. They are generally 6½ to 7½ inches. overall, with blades ranging from 2¾ to 3¼ inches.

Figure No. 21: Photos of the Original Gurkha Officer's "Patch" Knives (Souce: NMC)

During the muzzle-loading era, they are perfectly sized to cut away the excess patch material before loading. This hand-forged antique made before the 1800s has genuine horn handle. This knife was carried by 'Sardar' of Gurkha officers. It generally has 7¾ inches length with 3¼ inches blade. It has solid brass furniture.

A Visit to Museum

I personally regard the collections in the Nepal Museum Chhauni (NMC) as culturally and historically important artifacts. I was amazed to see particularly the collection of khukuris, which were used, by some historically important warriors and rulers of this nation. There are also some extraordinary collections of ancient firearms, weapons captured in various wars in Nepalese history, and also the uniforms worn by the earlier Nepalese Army at the NMC.

The Museum has a khukuri that is believed to have been used by the king of the then Gurkha kingdom Dravya Shah (1559–1570 AD). It is the oldest known khukuri found in the NMC. It is interesting to note that it has a heavy broad blade in '*Buduna*' style. The weapons that the

HM king Prithvi Narayan Shah (1723–1775 AD) and his troops once used are displayed at Nolche Darbar (Hanuman Dhoka), Basantapur, Kathmandu. One can also come across four different designs of khukuris, khukuri-bayonet fixed on rifle muzzle, *khundas, tarawars, kattaris,* Sundari rifle made in Sundarijal Kathmandu, Snider rifles manufactured by Yakubala in Nepal, and Gehendra rifle with long and unusual bayonets among others.

Regarding the history of arms and weapons, Perceval London in his book *Nepal Vol. II* writes, "Earlier than 1881 AD, Prithvi Narayan Shah enlisted the services of Francis Neville as head of the arsenal in Kathmandu. He is generally called a Frenchman, but as a matter of fact, was a half- caste. He was assisted by another semi- white man M. Dilbenee, and was making cannons on the river 'Tookihur', a mile south - east of Kathmandu". Tookihur is in fact the stream 'Tukucha' which runs through and under the old Royal Palace, where Singha Durbar is located today and finds its way into the Bagmati River beside Thapathali. "No magazines exist on its bank today. It is said that M. Dilbenee had with him a French colleague whose name was Vincent" writes London.

Visiting the museum one can view the bellied *buduna* style of khukuri, *tarawar* and *khunda* used by Surendra Pratap Shah. His armies seemed to have used two different styles of daggers.

Abhiman Singh Basnet (1744 – 1800 AD) is found to have used three different styles of khukuris as well as *khundas, tarawars* and *daggers.*

Kaji Kalu Pandey (Death 1757 AD) was an influential helping hand to the king Prithvi Narayan Shah. The khukuri that he once used is exceptionally heavy with a heavy wide blade with razor sharp even today, and is stamped "Pande". Pande must be quite a lusty gentleman to have been able to wield it. Pandey's troops used various types of khukuris that are more curved, heavy and 14 inch in length.

Similarly, there are khukuris designed in two different styles that were supposed to have been used by another historically important Royality Daljit Shah.

The museum also displays the personal khukuris of the Chautariya Fatte Jung Shah, another historically important Royal. Except his khukuri, there are other weapons such as *pommel, tarawar* and daggers.

The khukuri used by the Chautariya Krishna Shah, the Nephew of King Prithvi Narayan Shah and the *khukuri*, *tarawar* and dagger used by Kaji Nayan Singh Thapa and Sardar Ram Krishna Kunwar, the odd looking khukuri used by General Jagat Jung Rana, and fat bellied buduna style of khukuri used by Captain Sambhir Singh Rayamajhi are the artifacts that drew my curiosity.

Importantly, there is a *tarawar* with a revolver attached to its pommel, and it was supposed to have been used by General Jit Jung Rana.

The other historically significant weapons displayed at the Hanumandhoka Museum, Kathmandu include: some old Tibetan confiscated rifles during the Second Nepal-Tibet war in 1911 BS, old rhinoceros hide cannons which were locally made in Nepal, late King Birendra Bir Bikram Shah Dev's personal khukuri with the sun pattern designs, a Mercedes Benz car that was gifted by Adolf Hitler, once a German dictator known as 'Little Corporal' in World War I to King Tribhuwan, the old snider rifles manufactured by Yakubali in Nepal, the Gehendra rifles with long bayonet, the Sundari rifle made in Sundarijal arsenal in Nepal presently used only by the Gurujiuko Paltan Guards at the Hanuman Dhoka, some captured old Enfield Cavalry Rifles, and the Martini Henry Rifle with long and flat bayonet, old double barrel muskets among others.

Heroes Recognised

War seldom inspires love nor do enemies admire each other during or after a conflict. But of the 1814-16 war in the Himalayan foothills, Byron Farwell has written: "The love affair between the British and the doughty little Mongolian hill men had begun. It has lasted for more than 165 years". He wrote this almost forty years back, i.e 1980s.

In the course of an encounter early in the war, Lt. Frederick Young was captured by the Gurkhas. His soldiers from the plains of India had fled, leaving him surrounded by the hill men.

The captive British officer could have easily fled too, thought the Gurkhas.

They asked him, "Why did you not run away?"

The Englishman said, undaunted: "I have not come so far in order to run away".

The Gurkha soldiers were impressed, in their admiration, they said to him: "We could serve under men like you".

During his captivity, Young learned about his captors, their language, manners and customs. The more he learned, the more he admired the Gurkhas.

After his release, Lt. Young recommended the enlistment of his former enemies in the British Army. On 24 April 1815, the government authorized Young himself to undertake the task. He visited the POW camps and enlisted 3,000 volunteers. Thus, the first Gurkha battalion- the Sirmoor – came into being. In course of time, the Sirmoor became the 2nd King Edward VII's Own Gurkha Rifles.

During the same war, one Gurkha advanced towards the British lines, waving his hands as he approached. The British thought he had come out of the fort to surrender. His lower jaw had been shattered. He had come merely to seek surgical aid. After he had been patched up by an English surgeon, he asked for permission to rejoin his comrades in the fight against the British.

'I went there one man and came out three thousands', said Frederick Young.

'The pay of the Sepoy was a mere pittance of five rupees eight annas a month at that time' writes Ed. Smith in his book *Britain's Brigade of Gurkhas.*

Seeds of Friendship

The seeds of the long-lasting friendship between Britain and Nepal were not seeded in time of peace but in a short and bloody war.

In 1814, the Gurkha Armies invaded North India, occupying the hill stations of Mussoori, Simla and regional capital of Dehradun. The British administration in Delhi mounted a full invasion of Nepal with five battalion of Armies. Never mentioned, reported in the British history books, four of the Armies were totally repulsed and only the Fifth under General David Ochterlony, a great commander and tactician, succeeded in reaching to within sixty miles of Nepalese capital. At this point, the then King of Nepal, at the urging of his Prime Minister Bhimsen Thapa, sued for peace. The Treaty of Sugauli was signed on December 2, 1816. This treaty terminated hostilities, allowing the British to cross Nepal and commence trade with Tibet.

Lt. Frederick Young's first encounter with the Gorkhas was at the fort of Kalunga (Nalapani), where his commanding general, Gillespie, died in his arms.

At one stage, he was commanding some party of irregulars but fled when attacked by a Gorkhali force, leaving Young and a handful of officers at bay. From here onwards, a high degree of mutual respect existed between two countries. The highest praise the Gurkhas gave to their new comrades in arms were: "The English are as brave as lions; they are also splendid 'Bahadurs', very equal to us".

"Britain and Gorkhas had one quality noticeably in common, both were honest and incorruptible" writes, General Sir Francis Tucker.

Gurkhas formally commenced military service in the British Army from April 24, 1815. But to win the trust of the British Generals, they had to undergo many ordeals. The Third Anglo-Maratha War (1817- 18) and the Battle of Bharatpur (18-25-26) were the ordeals in which the Gurkhas proved their hardihood, loyalty and matchless valour to the British officers. Soon, the British considered the Gurkhas, by far the best soldiers in its army. Henceforth, they started extensively to mobilize the Gurkhas all over the world. Perhaps, there is no land on this planet where the Gurkhas have not fought for Britain. Their kike was the only closest friend in any hard times.

The Tripartite Agreement

The Tripartite Agreement signed on 9 November 1947 between and among Britain, India and Nepal makes provision for the recruitment of Gurkhas in the armies of Britain and India. Two days prior to this historical day, the Bipartite Agreement was signed between UK and India. According to the 1947 Treaty, Britain and India are not allowed to use Gurkha soldiers against internal affairs of any friendly nations of Nepal.

The Tripartite Agreement of 1947 had the following provision: "As Khukuri is the religious and national emblem of the Gurkhas forming also a part of the uniform of the Gurkha Army, the carrying of Khukuri by Gurkhas of all categories must not be banned in territories where the Gurkhas reside".

The Agreement was signed in triplicate in Kathmandu on 9 November 1947 by ACB Symon, for the Government of United Kingdom, Kanwar Daya Singh, Lt. Col., for the Government of India, and Padma Shumsher JBR for the Government of Nepal. The agreement in essence guarantees a provision for recruiting Gurkhas in the armies of the UK and India.

The Agreement states the following:

a. The Gurkhas and the brigade are integral part of the British Army.
b. The Gurkhas are not to be treated as mercenaries.
c. The welfare facilities for both the British and the Gurkhas will be the same.

However, the British Government was keen to obtain the services of some Gurkha regiments namely 2nd KEO GR, 6th QEO GR, 7th DEO and 10th PMO Gurkha Rifles. Later, the Queen's Gurkha Engineers, the Queen's Gurkha Signals and the Queen's Own Gurkha Transport Regiment were established in Malaysia.

Her Majesty Queen's Speech in Nepal

The speech of Her Majesty Queen Elizabeth II of the United Kingdom at a banquet given by His Majesty the late King Mahendra Bir Bikram Shah Dev on February 26, 1961 in Kathmandu goes:

"When my husband and I flew today over the terraced hills of Nepal into his beautiful valley of Kathmandu, I fulfilled a long cherished ambition, shared by countless numbers of my subjects for Nepal has often been in our thoughts. Since a child, the Sovereign's Gurkha Orderly Officers have been for me a constant reminder of your country. Many of my subjects, with indelible memories of service with and alongside the brave and cheerful men of Nepal, have longed to visit the homeland of such sturdy comrades and I am greatly privileged to be able to come here by your invitation.

It is fitting that I should once again express my gratitude for the staunch friendship and gallant support of your people and the Gurkha troops at many turning points of our history" (*This England* Golden Jubliee).

Queen's mother is reported to have told her daughter, who succeeded as Queen Elizabeth II, "Your work is the rent you pay for your room on earth".

Gurkhas are not Mercenaries

These facts given above make one clear that the Gurkhas aren't mercenaries. Had they been so, they wouldn't have been honored with the world famous awards such as the Victoria Cross of Britain and the Param Vir Chakras of India. Moreover, they wouldn't have been appointed as the King's Own Gurkha Officers (KOGOs) and the Queen's Own Gurkha Officers (QOGOs) at Buckingham Palace nor would they be appointed as the military ADCs to the President of India at New Delhi, Rastrapati Bhavan.

As a retired British Gurkha, deep down I believe that the Rana Prime Ministers of the then times were in fact the culprits for defaming Gurkhas. They were supposed to have pocketed 300,000 British Indian Rupees annually in the name of 200,000 Gurkhas who were sent to fight during the WW I on behalf of the Allies. In 1901, Chandra Shamsher classified the Ranas into three categories: A, B, and C. According to the new provision, only the 'A' class Ranas became eligible for ruling. But in the long run, it proved to be a seed of poison that brought end to their regime. In 1904, Chandra Shamsher arranged a term of peace during the trouble between British-India and Tibet, which resulted in opening a new trade route to Tibet via Kalimpong. It forced Nepal to give up its monopoly on trade so far carried on with Tibet. Chandra Shamsher was given the following titles by the British: Knight Brand Cross of the Most

Honourable Order of the Bath, Knight Grand Cross of the Most Exalted Order of the Star of India, Knight Grand Cross of the Most Distinguished Order of St. Michel and St. George, Knight Grand Cross of the Royal Victoria Order, Doctor of Civil Law, Honorary General, British Army and Honorary Colonel of the 4th Gurkha Rifles. This last was offered to him on a personal interest in supporting the 'Gurkha' regiments.

Such were the Ranas. But on the contrary, a quick look at the history of the Gurkha Rifles is enough for one to realise the rich tradition and heritage the Gurkhas have carved out in the British Army and also globally.

First Gurkha Rifles

The historically important First Gurkha Rifles was raised at Subathu, near Shimla, India by Lieutenant R. Rose in early 1815. Soldiers from among hill men, mainly the Kumaonis and the Garhwalis were selected to establish this regiment. The home cantonment of this regiment was later shifted to Dharamsala in the Kangra Valley. One should also mention here that this regiment's title was changed many times during the nineteenth century. From its new location at Dharmasala, the regiment moved for a short duration to Chakrauta, and then to Clement Town before resettling down at its original home at Subathu.

Second Gurkha Rifles (The Sirmoor Rifles)

This Gurkha Rifles was raised at Nahan, Sirmoor in early 1815 by Lieutenant Frederick Young. Soldiers from among hill men, mainly Kumaonis and Garhwalis who had fought under the Nepali General Amarsingh Thapa were selected to establish this regiment. These men, having been captured by the British during the war with Nepal, were held in camps in Sirmoor and Dehradun. Thereafter, Dehradun became the regiment's home until 1947.

Third Gurkha Rifles

This Gurkha Rifles was raised at Almora, Kumaon in early 1815. Hill men who had had interest to serve the British force were recruited in this regiment. The British General, Tuker opines that this regiment could be older than the Gurkha Brigade by a month or two. He claims that its birth took place in late 1814. Officially, Colonel Quhoun raised it a few days after the 1st and the 2nd Gurkha Rifles were established. Inevitably, its home was Almora. Then it was clubbed with the Ninth Gurkhas, and was located at Dehradun till 1975 when it finally moved to Varanasi in 1976.

Fourth Gurkha Rifles

As an extra regiment, the Fourth Gurkha Rifles was raised at Pithoragarh, Kumaon ten miles away from Nepal's western border in 1857 by Lieutenant D. Macintyre. He renamed it as the Forth Gurkha Regiment shortly afterwards. Its home was at Bakloh, close to the hill station of Dalhousie. Finally, being clubbed with the First Gurkha Rifles, it is now located at Subathu.

Fifth Gurkha Rifles

This Gurkha Rifles was raised in 1857 at Abbottabad in the northwest frontier province by Captain H. W. F. Boisragon. It was first named as the Hazara Gurkha Battalion. This regiment was one of the pre-1947 Indian brigades to belong to the Old Punjab Frontier Force. Its home was Abbottabad, and it became 'Royal' in 1923. Then it moved to Dehradun and had a separate training center for quite some time before it was clubbed with the Eight Gurkha Rifles Centre to be named 58th Gurkha Training Centre. It is located now at Shillong from 1976.

Sixth Gurkha Rifles

This Gurkha Rifles was one of the Old Assam Regiments whose title got changed many times. It was raised as the Cuttak Legion by Captain

Simon Fraser at Chaubiajganj in Cuttak, Orissa in 1817. It was renamed as the Assam Local Light Infantry in 1828. One year later it was again renamed as the 42nd Gurkha Light Infantry in 1886. Its home was later to be shifted to Abbottabad until it joined the British Army in 1947.

Seventh Gurkha Rifles

This Gurkha Rifles was one of the Burma Battalions raised in 1902 at Theyetmyo, Burma by Major E. Vansittart. Its home was eventually shifted to Quetta, Baluchistan, followed by the moves dictated by the Second World War that ended in 1947, the last time the regiment left India.

Eighth Gurkha Rifles

This Gurkha Rifles was another old Assam regiment raised at Sylhet in Assam in 1824 by Captain P. Dudgeon. It was first named as the Sylhet Local Battalion. It became the 44th Gurkha Light Infantry in 1886. The second battalion was raised in 1835, and it later became the 43rd Gurkha Light Infantry. General Tuker remarked that it was the second oldest battalion in the history of the British-Indian Gurkha Brigade. Its home was Shillong, Assam. It moved to Dehradun where 5th and 8th Gurkha had their separate centres. Finally, they merged at this place till it moved to Shillong in 1976 as part of the 58 GTC.

Ninth Gurkha Rifles

Ninth Gurkha Rifles was one of the old Bengal Battalions. It was raised at Fatehgarh in the United Province in 1817 by Major C. S. Fagan. It was first named as the Fatehgarh Levy. It became the Ninth Gurkha Rifles in 1901, and its home was later established at Dehradun. This regiment was the first one to recruit the higher caste Chhetris and Thakuris. Like the 3rd Gurkha, it also moved to Varanasi in 1976.

Tenth Princess Mary's Own (PMO) Gurkha Rifles

The history of the 10th PMO Gurkha Rifles goes back to 1766, the year it was raised as part of the East India Company's Madras Army. It was later conversed to a Gurkha unit that was raised by Lieutenant Colonel C. R. Macgreror from the Kubo Valley Police Battalion. The 10th Burma Regiment of Madras Infantry was designed here. It became the 10th Regiment (the 1st Burma Rifles) Madras Infantry in 1892. Then it was renamed as the 10th Regiment (the 1st Burma Gurkha Rifles), Madras Infantry 1896, and the 10th Gurkha Rifles in 1901. It is now being called as the 10th PMO Gurkha Rifles since 1950. Its home was first at Maymo, central Burma. The city Maymo was named after a British Colonel, May. The term 'myo' in Burmese language means a town. This regiment moved to Quetta before it further moved to the Kangra Valley after the Indian independence. Finally, the regiment became part of the British Gurkhas Army proper before moving to Malaysia in 1948.

I should also mention another historical fact that Britain started to rule Burma separately from 1937. After the WW II, a large number of Gurkhas, especially those who had served in the Second Burma Rifles started to settle their families in Burma. Later, they even started to work in Rubies Minefields too. They gradually acquired citizenship and have continued to serve in the Burmese Armies.

Eleventh Gurkha Rifles

This Gurkha Rifles Regiment was formed for a short time during the First World War. It was later reformed from the soldiers belonging to the 7th and the 10th Gurkha Rifles. These soldiers opted to remain in the Independent Indian Army. The bulk of their men were Kirantis from the eastern part of Nepal. They were raised at Palampur, and then were moved to Clement Town for a short while after which the centre moved to Ghanghora (Dehradun). In 1975, they moved to Katapahar, Darjeeling. Finding the place unsuitable for training at Katapahar, it finally moved to Lucknow in 1983, its present regimental centre.

Queen's Gurkha Engineer

The 67 Field Squadron Royal Engineers (RE), consisting of Gurkha infantrymen was attached to the RE raised at Kluang, Malay in October 1948.

The 69 Field Squadrom RE was raised at Kluang, Malay in August 1950. Similarly, the RHQ 50 Field Engineer Regiment was raised in Hong Kong in June 1951. The Queen's Gurkha was formed as the Gurkha Engineers, part of the Brigade of Gurkhas, on 28 September 1955. The present designation was granted on 21 April 1977.

The 70 Gurkha Field Park Squadron was raised at Sungei Besi on 1 April 1960, and was disbanded in Singapore on 31 July 1971.

The 70 Gurkha Field Squadron that was raised on Sungei Besi on 1st April 1961 was later disbanded in Hong Kong on 17 August 1968.

Her Majestry Queen Elizabeth approved the title of the Queen's Gurkha Engineers on 21st April 1977.

The 69 Gurkha Independent Field Squadron was reformed on 1st March 1981 in Hong Kong and the Support Squadron designated 70 Support Squadron on 6 August 1982.

Current Organizations under Queen's Gurkha Engineers since 1983

Regimental Headquarters, the Queen's Gurkha Engineers
67 Gurkha Field Squadron
68 Gurkha Field Squadran
69 Gurkha Independent Field Squadron
70 Support Squadron

Queen's Gurkha Signals

It was raised at Kuala Lumpur in 1948 by Major AC Cox. Royal Signals from Gurkha Soldiers of each of the eight battalions of the then New Brigade of Gurkhas were enlisted in this regiment. This, together with re-enlisted ex-servicemen of the Indian Army, British officers and soldiers of

Royal Signals was formed as the Gurkha Signals Training and Holding Wing in early 1949.

The first elements were designated as the Royal Signals Gurkha. In 1952, this was changed to the Gurkha Royal Signals. The present badge was granted to this regiment on 23 September 1954, which is now the official regimental birthday. The regimental title became the Gurkha Signals in 1955, and Her Majesty Queen Elizabeth II approved the title of Queen's Gurkha Signals on 21st April 1977.

Affiliated Corps: Royal Corps of Signals

Affiliated Regiment: 32nd (Scottish) Signal Regiment (volunteers)
Official Association with 6 Signal Regiment Royal Australian Corps of Signals.

Queen's Own Gurkha Transport has the following history.

The 28 Company Gurkha ASC was formed in July 1858.
The 31 Company Gurkha ASC was formed in July 1959.
The 30 Company Gurkha ASC was formed in July 1958 and was disbanded in December 1968.
The 34 Company Gurkha ASC formed in July 1959 and was re-designated to the Gurkha All Arms MT Training Wing GTR on 1 July, 1970. It was further re-designated as the Gurkha MT School and was absorbed by the 31 Squadron GTR on 19 July 1971.

The Headquarters of the Gurkha ASC 17 Gurkha Division was formed on 1st July 1960.

And, finally the Gurkha Army Service Corps was re-designated as the Gurkha Transport Regiment in 1965 and the 415 Maritime Troop RCT were included in the regimental establishment from 1st January 1979.

Current Organizations of Queen's Own Gorkha Transport since 1983:

Regimental Headquarters Gurkha Transport Regiment
HQ Squadron (includes 415 Maritime Troop RCT)
28 Squadron
29 Squadron RCT
31 Squadron

The Gurkha Contingent of the Republic of Singapore Police Force:

The Gurkha Contingent of the Republic of Singapore Police Force was formed in 1949 at the conclusion of WW II when the Government of Singapore sought a unit to replace an existing Sikh's unit that had disintegrated under the Japanese occupation. There was no agreement between Singapore or Brunei and the Ministry of Foreign Affairs of Nepal. Strangely, the Gurkhas are not recruited directly by the Singapore Government, but through the British Army, which receives "commission" for the job.

The Pasokan Simpanan Gurkhas (Gurkha Reserve Unit): It was established on 25 February, 1974 in Brunei.

The United British Gurkha Ex-Servicemen Association, Nepal (UBGEAN) was established in 2005.

Insignia

Khukuri now appears in badges of all Gurkha Regiments and Units, with an ingenious variety of arrangement. With the crossed 'Kikes' their cutting edges outwards and Her Majesty Queen Elizabeth II's crown centre top is the most prestigious badge that the Gurkhas adore so much in their life. There is a specific meaning behind each Regimental Badge that incorporate Kike. For Example: Of my own Regiment, the 10th

Princess Mary's Own Gurkha Rifles has badge that has bugle horn stringed ot interlaced with a Kike fessewise the blade to the sinister. Above the Kike has the inscription CI GCVO GBE ED RRC DCL HRH Princess Mary (The Princess Royal) and below it the numeral 10.

Though ancient weapons such as bow and spear have become antic pieces subject of museumization, the Kirantis' Kike has acquired greater forms and global recognition. Kike is still used by Kirantis in their daily works in the hills and mountains. Many can still be seen carrying it wherever they go in hill and mountains.

Khukuri was also proudly displayed along with other national icons on all notes of Ruppess issued by the Nepal Government. But the government in recent times has stopped printing the insignia of khukuri on all paper notes. As a veteran Gurkha, I request Nepali government to reconsider this very decision. Khukuri has remained in use since time immemorial. Rulers in this country have risen and fallen. But khukuri or kike has remained as a source of power and protection for people. Let's cherish this fact.

In World War II, it was also featured in the shoulder flash of 43 Gurkha Lorried Infantry Brigade, and after the War, it appeared in the formation signs of many Gurkha Units including 18th, 48th, 51st, 63rd, 99th Brigades, the Gurkha Field Force in the Far East and by 28 Squadron, the Gurkha Transport Regiment in the Gulf War.

The following Gurkhas Regimental Head Dress Badges Incorporating "Kikes' are shown below-

Figure No. 22: The Gurkha regimental badges incorporating khukuri.

1. 1st King George V's Own Gurkha Rifles (The Malaun Regiment) 2. 2nd King Edward VII's Own Gurkha Rifles (The Sirmoor Rifles) 3. 3rd Queen Alexandra's Own Gurkha Rifles 4. 4th Prince of Wale's Own Gurkha Rifles 5. 5th Royal Gurkha Rifles (Frontier Force) 6. 6th Queen Elizabeth's Own Gurkha Rifles 7. 7th Duke of Edinburgh's Own Gurkha Rifles 8. 8th Gurkha Rifles 9. 9th Gurkha Rifles 10. 10th Princess Mary's Own Gurkha Rifles 11. The Queen's Gurkha Engineers 12. Queen's Gurkha Signals 13. The Queen's Own Gurkha Transport Regiment 14. Royal Gurkha Rifles 15. 11th Gurkha Rifles 16. Gurkha Reserve Unit Brunei 17. The United British Gurkhas Ex-Servicement Association, Nepal

Procedures for Recruitment

Himalayan hills were once and still are traditional rang rooting ground for one to get recruited in the British Gurkha Army.

Procedures for getting recruited are as given below:

a) Medical assessment

One needs to get his chest X-rayed to ensure that the would-be Gurkha is not a patient of tuberculosis and other conditions. He or she should also get his or her hearing and other deformities checked. After the medical exams, one and half a mile needs to be completed by a candidate within 9 min. 40 seconds.

b) Physical test

As part of the physical test, a candidate needs to carry 25 kg sand in a traditional Nepalese 'Doko', a cone shaped bamboo basket. It is carried on the back with a 'Namlo' strap around their forehead. A test like this is just as much about mental as physical one, experts say. Other tests a candidate has to go through include push and press up in said timings.

c) Documentations

Candidates should get their educational certificate submitted. He or she should have minimum a SLC or School Leaving Certificate degree. Importantly, he or she should also take a written test exam in the English medium.

d) The final selection

The British Gurkhas and the Gurkha Contingent of the Republic of Singapore Police Force have their own orders of merit list that candidates have to fulfill. For example, candidates should be between 18 and 21 years of age.

As a senior retired Gurkha soldier, I advise that close quarter combat fighting techniques with khukuri must be taught to all Gurkha soldiers.

Global Gurkhas

Gurkhas have gone global right from the time after the then Gurkha's rulers unified Nepal. Their name and fame travelled in her neighboring countries in addition to the British Army. In this process, they have served a number of other countries as well.

1. Gurkhas in Bhutan

It has been a tradition among Nepali youths to serve in foreign armies since as early as the seventeenth century. In 1624, Bhutanese King Sabdung Nawang Namgel met with Gurkha King Ram Shah and established a friendly relationship with him. At the request of this king, a large group from Nepal's martial races, like Magars and Gurungs, went to Bhutan and settled there under the leadership of Bishun Thapamagar. The King of Bhutan again visited Gurkha in 1640. This time, he also visited Kathmandu and requested another large group of Nepali people to immigrate to Bhutan. They settled in the border area of this country. Most of these people served Bhutan as soldiers, and some of them were entrusted with border defense. Later, Bhimsen Thapa provided a number of senior Nepali army officers to train the Bhutanese Army besides offering a considerable quantity of rifles and ammunition to Bhutan.

2. Gurkhas in Sikkim

Sikkim, once a separate country and now India, adjoins Nepal. The population of Sikkim is made up of Lepchas, Bhutias, and people of Nepali origin - Rais, Limbus, Magars, Gurungs and Chhetris. They have been contributing to the defense of Sikkim. In 1868, Nepalis such as Khasa Dewan, Phudong Lama and Sidkeyoung Namgyal got a large number of Nepalese to settle in Sikkim in order to protect its sovereignty, excavate its mines and mint coins. There were about 50,000 Nepalis in Sikkim between 1887 and 1908. Before Sikkim was merged into India, many Gurkha soldiers such as Captain Rolan Chhetri, Captain Lal Bahadur Limbu and Rifleman Basanta Kumar Chhetri along with the Magar rank and file had served the palace of King Chogyal.

3. Gurkhas in the Punjab

King Ranjeet Singh of Punjab was highly impressed by the outstanding valour of the Gurkhas displayed during the Anglo-Nepal war of 1814-16. Owing to this reason, he gave an honoured place to Gurkhas in his army. Bhupal Singh Thapa, brother of the legendary soldier Amar Singh Thapa, served this army. Through his influence, other Gurkhas joined that army. In fact, Amar Singh Thapa himself joined Ranjeet Singh's army, so did another legendary Nepali fighter Balabhadra Kunwar. Kunwar was paid Rs. 10 a day, and later he became one of the top Generals of the Army as well as a confidant to the King.

There were Gurkhas serving as infantry soldiers in Ranjeet Singh's army as early as 1809. However, it is still unclear when exactly did the Gurkhas begin to join Singh's army. A war had ensued after the annexation of Peshawar of Pakistan by Punjab in 1822. The Gurkhas fought alongside Punjabis and displayed their finest warring qualities in that war. In the course of the fighting, both Balabhadra Kunwar and Amar Singh Thapa were killed. It proves that the Gurkhas were recruited in the Punjab prior to the British India.

4. Gurkhas in India

After pulling out of India, Britain left six Gurkha regiments behind. On 01 January 1948, India raised one more Gurkha regiment (the 11th Gurkha Rifles). More than 50,000 Gurkhas (Nepalis and Indi's domiciled Nepalis) are currently serving in the Indian Army. The Gurkhas have been engaged in all wars fought by India. The Gurkhas serving in the Indian Army have absolute equality with their Indian counterparts, whereas, there still remains blatant discrimination even after a two decade long marathon movement in the UK.

It should be noted that the legendary soldiers Balabhadra Kunwar and Amar Singh Thapa deserted the Army and joined Punjabi Army (a former adversary). It is a serious crime to desert the Army during wartime, that too, by such legendary Generals. However, their bravery had been recognized by the state and they were included in the national pantheon. To our dismay, the 13 Gurkhas who won Victoria Crosses, and who fought all over the world during the First and the Second World Wars, and who had been recognized as the bravest of the brave by the world, have not yet been added to the national pantheon in Nepal. It is high time, the leaders and thinkers realised this. In the present world, there might be very few countries that allow their citizens to join foreign armies. Therefore, it is also high time we made discourse and analyzed its good and bad sides and reach decision suitable to the times.

5. Gurkhas in Burma

After the complete occupation of Burma in 1886, the British Government recruited 2,240 soldiers, most of them Gurkhas. After receiving some training, they were organized into various forces: Burma Frontier Force, Burma Military Police, Rangoon Armed Police and others. In addition, there was a company of Gurkhas in the 17th Burma Regiment. Moreover, a number of Gurkhas worked for the military police, others for the transport company and still others worked for various forces. In 1937, Britain started to rule Burma separately from the British India. Then, they set up the organization, the Burma Rifles. The Gurkhas were included in these regiments. The Second Burma Rifles consisted of both Gurkhas and Burmese. This regiment gallantly fought against the Japanese during the Second World War. After World War II, a large number of Gurkhas who served in the Second Burma Rifles settled in Burma with their families. They later acquired citizenship, and continued to serve in the Burmese Army as do their children.

"Hanthawaddy" – during the British colonial rule, the term "Hanthawaddy" district was used to cover the north central region of Burma. This area includes the ancient cities of Pagan and Mandalay, and the hill city of Maymyo. There are several Gurkha towns in the Hanthawaddy districts. The name "Hanthawaddy Bando" was selected to honor the martial discipline of the Gurkha warriors of the 10th GR. After their service, hundreds of them retired and established their residence in that region.

6. Gurkhas in China

In 1906, a Chinese diplomat based in Lhasa made a request to the Nepali Consul to begin recruiting Gurkhas into the Chinese Army. But the Nepali Government turned it down. Therefore, the Chinese Government recruited nine half-breed Nepalis (born of a Nepali father and Tibetan mother) in 1909. The Nepali Consul based then in Lhasa mentioned their names in a letter dispatched to Nepal's Prime Minister Chandra Shamsher. A group of Gurkhas including Corporal Man Bahadur Limbu and Corporal Asha Bir Rai deserted the British Army, the 10th Gurkha Rifles, and joined the Chinese Army in Yunnan, where they served as drill instructors. When news of this reached the British, the Commander in Chief initiated diplomatic efforts to stop the practice of Gurkhas joining the Chinese Army, writes Ram Narayan Kandangwa.

Gorkha's Service with American Units

My brother Man Bahadur Rai 3/10th GR recalls the service with American Units. Thousands of Gurkha soldiers were recruited to serve with various American Military groups during the Burma Campaign in the World War II.

American Unit	**Gurkha Service**
1. US OSS-101	Protected HQs
2. US Merrill's Marauders	Served as scout and orderlies
3. US Kachin Rangers	Served as scouts and rescue teams
4. US Signal Corps	Protected signal stations
5. 1st US Air Commandos	Protected air bases
6. US Flying Tigers	Protected air bases
7. B-29 Bomb Wing Group	Protected air bases
8. B-25 Bomb Wing Group	Protected air bases
9. C-47 Air Supply Group	Protected air bases
10. US LP Evacuation Planes	Protected air bases
11. US Radar Stations	Protected radar stations

12. US General Stilwell	Protected HQ
13. US Chinese Forces in N. India	Protected HQs
14. American Field Ambulance	Protected motor depots
15. American Field Hospitals	Protected perimeters
16. American Supply Depots	Protected compounds
17. Stilwell Road	Protected construction sites
18. American Motor Depots	Protected compounds
19. American Liaison Officers and many other duties	Personal guards

The Gurkhas served the Americans with loyalty and great dedication. They respected the Americans for their single-minded determination to complete their assigned mission and for their courage under fire. They loved the Americans for their genuine kindness, friendliness and equal treatment of men regardless of race.

In 1964, Dr. M. Gyi formed the American Bando Association, (ABA) as a private non-profit martial arts organization to honor and pay tribute to the veterans of the Burma Campaign. The term 'Bando' is an ancient Burma word for "martial discipline" and "combative arts". Crossed khukuri is the symbol of the ABA.

'Band Shay' uses a variety of weapons, the key weapon being the khukuri, but it includes others such as a stick, cane of any sizes and shapes or any other implements like a roal.

Global Gurkha Warriors with Kike

The countries where the Gurkha warriors fought with their kike during the WW I & II and subsequent wars under the British Army are as follow:

1. Afghanistan	1879, 1919	With Imperial Britain
2. Bhutan	1865	With Imperial Britain
3. Belgium	1914	With Imperial Britain
4. Burma	1941, 1945	With Imperial Britain
5. Brunei	1963	With Britain
6. Bangladesh	1971	With India
7. Belize	1978	With Britain
8. China	1900	With Imperial Britain
9. Caucasus	1919	With Imperial Britain
10. Congo	1961, 1962	With Britain
11. Cyprus	1978	With Britain (I served this war)
12. Egypt	1942	With Imperial Britain
13. England	1962	With Britain
14. France	1914	With Imperial Britain
15. Falklands	1982	With Britain
16. Gallipoli	1915	With Imperial Britain
17. Greece	1919, 1940	With Imperial Britain
18. Gulf	1991	With Britain
19. Hong Kong	1970	With Britain (I served this war)
20. India	1815	With Imperial Britain
21. Iraq	1915, 1920	With Imperial Britain
22. Iran	1918, 1941	With Imperial Britain
23. Italy	1944	With Imperial Britain
24. Japan	1946	With Imperial Britain
25. Java	1946	With Imperial Britain
26. Kashmir	1962, 1965, 1971	With India
27. Libya	1942	With Imperial Britain
28. Malaya/sia	1875, 1941, 1948	With Imperial Britain
29. Maldives	1988	With India
30. NEFA	1962	With India
31. Palestine	1918	With Imperial Britain
32. Soviet Central Asia	1920	With Imperial Britain
33. Singapore	1960	With Britain (I served this war)
34. Sri Lanka	1987	With India
35. Tibet	1903	With Imperial Britain
36. Tunisia	1944	With Imperial Britain

The World War II was truly global. Sixty-one countries participated in it. They held 75 percent of the world's population. Over 110 million people were drafted into the armed forces.

During the World War II (1939- 1945), the whole of Nepalese army was placed at the disposal of the British crown. Eight Nepalese regiments were sent to India for internal security duties and for operations on the North West Frontier. Later, a brigade was sent to Burma, and it fought with particular distinctions, especially in the battle of Imphal.

A large sum of money for weapons and equipment, including money for the provision of fighter - aircraft was handed over as gifts from Nepal. Similarly, during the blaze, for the relief of victims in the Dockland Area, a considerable sum of money was also donated to the then Mayor of London. And, an equally generous response was made to a variety of appeals for aid.

The spirit of this friendship can best be illustrated by the reply made by the Prime Minister of Nepal to the British minister in Kathmandu. After the fall of France, when Britain stood alone, an additional 20 battalions for the Gurkha Brigade was sought to recruit. This permission was readily granted by the then Prime Minister who remarked, "Does a friend desert a friend in the time of need? If you win, we win with you. If you loose, we loose with you". Naturally, the same quantities of kikes were issued to them also.

The past histories show that Gorkhali warriors with their kike have defeated their foes equipped with tanks, cannons, long-range missiles and guns. They have made glorious victories. Ever since kike has symbolized courage, bravery, security, and honor in the past and present the Nepal Army, the Indian Gurkha Armies and the Royal Gurkha Rifles of the British Armies.

In the First World War, the Gurkha battalions were to be found wherever the fighting was more deadly - in France, at Gallipoli, in Palestine and Mesopotamia. For the Gurkha regiments, there was no peacetime in its literal sense between the two World Wars, for they constantly stood watch on India's North-West Frontier. And, in the Second World War, Nepal sent nearly a quarter of a million men to fight against Germany and Japan. The end of the War against Japan brought peace to the Far East.

'Each little Gurkha might be worth his full weight in gold at Gallipoli', wrote the General, Sir Ian Hamilton, in March 1915. During the World War I and II, Nepal suffered the highest per capita combat casualties of all the nations in the world, writes Dr. Mary Des Chene. The Gurkhas have been on our side in every war fought by the Great Britain, not least in the recent Falklands conflict. The late Brigadier E. D. Smith DSO. MBE 2nd GR, and Colonel of the Brigade of Gurkhas from 09/ 01/1971 to 12/ 07/1971 wrote:

"As I write these last words, my thought returns to you who were my comrades, the stubborn and indomitable peasants of Nepal. Once more, I hear the laughter with which you greeted every hardship. Once more, I see you in your bivouac or about your fires on force march or in the trenches, now shivering with wet and cold, now scorched by a pitiless and burning sun. Uncomplaining, you endure hunger and thirst and wounds, and at last your unwavering lines disappear into the smoke and wrath of battle. Bravest of the brave, most generous of the generous, never had a country more faithful friends than you".

Sir Ralph Turner MC in the Preface to the Dictionary of Nepali Language writes: There is probably no place on earth where Gorkha's sturdy feet have not stepped upon, wherever there was a threat of war.

Gorkhalis have been deployed in the front lines in every British Military Campaign: the Maratha and Sikh's War; the Sepoy Mutiny; the Afghan War; the Boxer Rebellion in China; during the World Wars, virtually in every theatre of battle; in the post-second World War insurgencies in the Southeast Asia, and most recently, in the Falklands, the Gulf, Kosovo, Afghanistan and now in Iraq. The Gurkhas were also deployed in curbing the Indian nationalist uprisings in the last quarter of the nineteenth and the early twentieth centuries.

After the Second World War, those Gorkhali soldiers who served for the British Crown with their khukuris have guarded the shred of Great Britain from the Buckingham Palace to the oilfield of Brunei. The Malaysia emergency might not have been controlled; the Indonesian confrontation might not have been overcome, the Falkland Islands War in 1982 would not have turned into the favour of the UK, and also it would have been very difficult to maintain law and order during the Cultural Revolution in Hong Kong till 1999.

During the WW II, over 150,000 Gurkhali soldiers not only fought with their rifles, bullets and bayonets, but also used their khukuris to fight with the Allied Forces against the German and the Japanese Imperial Forces. The Gorkhali with khukuri became the most feared enemy to them during the Wars.

The Gurkhas gave the most that men can give life itself for god, king and country for loved ones at home, and the United Kingdom for the sacred cause of justice and the freedom of the world. Images of khukuri are synonymous with the images of Gorkhas. This reality can be realised not only in hamlets of the Himalayan region but also right in the heart of the western metropolis. For example, while walking in the beautifully-kept British War Cemeteries of the Western Front, one is reminded of a stocky breed of the fighting man from a distant land as he sees the device of crossed 'khukuris' (kukri), the curved knives of the Gurkhas on a headstone. Many accounts of the Gurkha's prowess with this weapon occur in their Regimental histories, yet little attention has been given to the weapon itself', mentions the National Army Museum, London.

Khukuris and Modern Law

Khukuri has become a thing of concern globally. Should it be allowed to carry for the public? It has triggered debates in the UK now. The key text of a directive issued by the Brigade of Gurkhas on 6 November 1996 mentions the following things:

1. Khukuris

   There has been considerable public concern recently in the United Kingdom over the issue of "combat kukris" and prominent calls for legislation to make the sale or possession of them illegal. Whilst the legislators wrangle over the difference of 9" long kitchen knife and a 9" serrated edged bowie knife with camouflaged handle and incorporated knuckle duster, it is quite clear that the khukuri by any British definition is a combat knife.

2. Khukuri is possibly the single most readily identifiable image, associated with the Gurkha, and it is unfortunate that khukuri features prominently in any display of knives handled in or seized by the police. However, the 1947 Tripartite Agreement contains specific reference to khukuris, their importance as national and religious symbols and the fact that their carriage by the Gurkha servicemen should not be banned. It is important therefore, that the status of khukuri is not demanded and that no opportunity is given to question the right of the Gurkhas to carry khukuri, i. e. we must put our house in order in advance of any new laws.
3. Addresses are asked to note the following principles for passion and carriage of khukuris:-

    a. Khukuris are issued to the Gurkha soldiers on a scale of one per man for ceremonial use.
    b. The Gurkha may also choose to carry a non- service issue khukuri for field use.
    c. Khukuri may only be taken out of a Gurkha barracks by the Gurkha personnel on duty, such as parades, field training, demonstrations and exhibitions.
    d. Service personnel and Sunars employed by units are forbidden to sell or trade khukuris. Gurkha units are prohibited from importing, selling or giving away khukuris whether they are obtained from ordnance or other commercial sources.
    e. The formal presentation of khukuris is to be restricted to individuals known to the unit and who are of a reliable character, etc.; or to a responsible organization with close ties to the brigade e.g. Hard District Council.

4. You will appreciate that any attempt to set out rules, which are workable and cover all eventualities is extremely difficult. It is also questionable whether restrictions on sales would apply in Brunei. We

may in future need to address the problem of importing and exporting khukuris from and to Nepal.

5. For the Gurkha Museum and the Gurkha Welfare Trust as far as I am aware, the museum shop no longer trades in khukuri although their sale was always a reliable source of income. You may appreciate the adverse publicity, which would follow from any khukuri incident in which if the khukuri involved is traced back to the sale.

In late 20th and early 21st century, ninety percent of the world's khukuris were bought up by an individual in the United Kingdom. The article written by highly respectable martial art instructor master Simon Hengle also happens to be one of the finest kike or khukuri-fighting practitioners outside the Brigade of Gorkhas to this day.

Thus, the Gurkhas and their khukuris have become global now. They have served the Empire. And, while serving the Empire, they have served humanity. Their khukuris have become global. Nepal should evoke this heritage the Gurkhas and their khukuris have carved out over the centuries. Khukuri or kike handicraft industries can become a very strong source of income for Nepal, an opportunity to grab and explore.

Chapter V

Facts and Fictions about Gorkhas

The world famous Colombian writer Gabriel Garcia Marquez described the Gurkhas as "decapitators who beheaded Argentinean soldiers with their scimitars, at the rate of one every seven seconds. They held to serve by the hair and looped off the ears." Garcia Marquez provided no evidence to support this claim, which was described to an extreme case of war jitters in Latin America; certainly Marquez's report contradicted the account of Falklands veterans, who spent a bonus leave in Nepal.

Marquez was no doubt the most influential writer. He portrayed the Gurkhas as deadly and dangerous inhuman merceneries. Such portrayal stands far leass effective than the romantic and heroic portrayal the popular writing has made about the Gurkhas. Long back, i.e. at the end of the nineteenth century, *The Reverend Wood*, a tour guide text book admired the Gurkha and their khukuri in the following manner: "Indeed, the Gurkhas are small in their height. Their hands, like those of all Indian races, are very delicate, and are short. The point of khukuri is as sharp as a needle, so that the weapon answers equally for cutting or stabbing. In consequence of the great thickness of the metal, the blade is exceedingly heavy. It may be imagined that a blow from such a weapon as this must be very terrible one. The very weight of the blade would drive it half through a man's arm if it were only allowed to fall from a little height".

No doubt the Gurkhas hold knowledge about how to strike target with khukuris. They also know how to sliece enemy's parts of body. Since they are culturally trained to handle it, they feel connected to this weapon. They use it daily. They know how to invent strikes according to the target and situation. This allows them to control their mind, the breathing pattern and the footstep. But when the publication like *The Rev. Wood* started projecting the Gurkhas from the perspective of the British Colonialism, the image of the Gurkhas become articulate: "Until our men learned this mode of attack, they were greatly discomfited by their little opponents, who got under their weapons, cutting or slashing with knives as sharp as razors and often escaping unhurt from the midst of bayonets. They would also dash under the bellies of officers' horses, rip them open with one blow of khukuri and aim another at the leg of officer as he and his horse fell together". Such was the popular assumptions about the Gurkhas. What we can say about this mode of image production of the Gurkhas is that most bad rumours come true. Most good ones don't.

On the other hand, the soldiers and the ambulance drivers from the battelfied, who were also the cofighters of the Gurkhas, their reports and writings provided authenticity to the already popular images of the Gurkhas. It made people to regard facts and fictions about the Gukhas together:

"The Gurkha!"

In far Nepal, there's a breed of a race,
Sturdy, strong and brave,
For many years they have fought for us,
Since first their allegiance they gave,
Their battle honours bear cut their fame
Their glory from shore to shore,
With a Gurkha fighting besides him,
No soldier could ask for more,
What a debt we owe him,
This man who from a far,

Has fought besides us with honour,
And has made us what we are,
On foreign fields, he's spilled his blood,
And from cradle to the grave,
We have no doubt our comrade
Is the bravest of the brave.
He's honest, willing and cheerful,
Never without a grin,
And if you need a friend to help,
You can always count on him.
He'll fight on when the task seems hopeless,
And when the situation's grave,
You'll know, then why they call him,
The bravest of the brave.

(Composed by Majorie and Royde Vondre)

Poets and writers have pointed out many multitudes of Gurkha warriors. They honestly believe that the Gurkhas gave the most that a man can give life itself to God, King and Country for the loved one's at home and the United Kingdom for the sacred cause of justice and freedom of the world during the Great War of 1914–1918. Another poem below is a further proof:

A Prayer for the Gurkhas

O God, who in the Gurkha, has given to mankind
A race exceptional in courage and devotion,
Resplendent in its cheerfulness.
We, who owe them so much,
Ask Your special blessing on them,
Their families and their land.
Grant us Your Grace to be loyal to their
Best interests, as they have been to ours in the past
(From *The Bugle and Kukri)*

A number of tall, wild and colourful tales about khukuri usually come from people other than the Gurkhas themselves. Interesting anecdotes and stories about khukuri and the Gurkhas told by non-Gurkhas abound. One story has it that once a Gurkha unsheathes his khukuri, he will not keep it back until he has drawn his enemy's blood, failing which he pacifies himself by cutting his own finger.

Some written views expressed in various articles are those of respective authors, journalists, editors and few former distinguished British and Indian officers who once served one or the other Gurkha regiments:

An International Force comprising the British and the German among other troops were stationed in Shanghai during the so-called Boxer Rebellion in 1900. Count Waltersee, the German Commander of the Force mentions two Gurkha riflemen, who had been posted by the British as orderlies to their officers as his staffs in a dispatch to Berlin. He praises the way the Gurkhas demonstrated a skillful use of khukuri in cutting wires connecting numerous land mines. This enabled the Germans to advance. The Commander then requested the German government to award the Gurkhas with medals for their feat.

The Amrita Bazar Patrika, a daily published from Calcutta dated 9 November 1914 carried the following story: "France is called the land of fashion. Fashion in any field begins from here. Even in the First World War in 1914, a fashion was started here, which reflected the glory of the Gurkhas. Fascinated by the valor of Gurkhas in battle, the French soldiers, citizens and especially young women were very interested in seeing khukuri carried by Gurkhas. But when it was rumored that khukuri may not be seen at all times, and that once it is unsheathed, it had to draw blood, the French stopped their demand to see it.

But a few people in their irrepressible wish to see Khukuri had the tip of their fingers cut. This was soon known everywhere. And then, soldiers and citizens and even young men were seen in the streets of Paris with "cut fingers". The fashion became so widespread among young women that they held up their bandaged finger as a proof of having seen Gurkhali weapon".

In a true story told by the Field Marshal, Sir William Slim in his outstanding account of the Burma Campaign in his world famous book *Defeat into Victory*, he recounts:

> 'After this little excitement, the tank having, to our relief moved again to a flank, we watched the final stages of the action. The fire of Brens and rifles swelled in volume; the tank's gun thudded away. Suddenly three Gurkhas sprang up simultaneously and dashed forward. One fell, but the other covered the few yards to the bunkers and thrust Tommy guns through loopholes. ...They swarmed around the bunkers and for a moment all firing ceased. .... Within ten minutes, having made sure no Japanese remained alive in the bunkers, the two platoons of Gurkhas and their Indian-manned tank moved on to their next assignment, which would not be far away. A rear party appeared, attended to their own casualties, and dragged out the enemy bodies to search them for papers and identifications. It was all very business-like. If I have given more space to this one incident, that was being repeated in twenty places in the battle, than I have to much more important actions. . . . The men who carried it out were from a Gurkha regiment of which I have the honour to be Colonel' (450).

In early 1944, unorthodox guerrilla fighters Brigadier Orde Wingate's Chindits and, in particulars, General Stilwell's 'Long Range Penetration Group' (US Army) known as 'Merrill's Marauders' became a thorn for the Japanese in their occupation of Burma. Inserted and supplied by air, they operated behind the lines without relief or reserves and with minimal support. A precarious situation existed. They had to live in the jungle and keep on moving all the time. Khukuri provided itself a great help, beyond doubt easily surpassing knives, bayonets, parangs and general issue machetes.

A well-known Indian Army man Hashim Khairullah writes: "These Gurkha regiments have proved to be a great asset to India's army and have distinguished themselves by their valour and steadfastness in all of

India's post- independence wars. The valour which has been inherited through the centuries in the lofty mountains of Nepal will continue to endure in far flung battlefields".

Gurkhas are not supermen, but they are certainly among the finest soldiers in the world, writes David Stelle.

Similarly, Lt. General Sir Brian Horrocks once wrote: 'On one occasion, we were suffering from an intermittent long range enemy artillery bombardment which interfered with the RAF's use of an important forward airfield. I asked Tuker whether a fighting patrol of Gurkhas could slip through the enemy lines during the night and eliminate this nuisance, and he agreed. Apparently, prior to the patrol's departure, there had been an argument as to whether a British officer was capable of cutting off a German's head with one blow from a khukuri, with which the Gurkhas are expert. The raid was a complete success, and from then on, the airfield was used by our aircraft without interference. A couple of days later, I was invited to the Gurkha Officer's Mess and presented with a khukuri with which I was assured the British Officer in charge of the patrol had successfully dealt with the German sentry in charge of the gun – with one blow'.

This same khukuri is now hanging up in my study room though it has never been regarded with much favor by the female members of my family.

I believe that the Gurkhas rank at the top of the world's all most formidable fighting men. Apart from the Big Boy and the Fat man, the atomic superstars of the First and the Second World Wars, three weapons should top billing as the most famous: the Luft Waffer's July-87 Stuka, the US Armed Forces' quarter-ton jeep, and the khukuris of the British Gurkha Brigade.

A great boulder on the slope of Monte Cassino in Italy bears a crossed Khukuri badge of the 9th Gurkha Rifles. "For, in this hilly country that seemed made to test their particular skills – the night attack, the surprises from the rear, the small patrols, the hand-to-hand combat – the Gurkhas had given a demonstration of their skill in what Churchill himself had described as "a brilliant feat of arms", writes Horrocks.

There is another story about the Gurkhas. The war in the Middle East was not over for all the Gurkha Battalions. They were ordered to leave Mesopotamia for Persia. Here, the Persians' sympathy undoubtedly still lay with the Germans. Both the Battalions and the Bolsheviks took part in operations against a local guerrilla leader, one Kuchik Khan, whose force was called 'the Jangalis'. Fortunately, an early brush with these gentlemen in late July 1918, found the Gurkhas getting well home into the enemy, which 'gave them a wholesome dread of the Gurkha Kukri, the fame of which weapon spread through the countryside'. Both battalions had several spirited actions against the 'Jangalis' before operations in Persia came to the end, nearly three years after the war in Europe had ended.

Not very different is the story (set variously in China, Italy, Burma and the North West Frontier) of the Gurkha coming suddenly on an enemy solider. Naturally, he struck the first decapitating blow. "Yah, missed!" said the enemy. "Try shaking your head", came the reply.

The Gurkhas are highly amused by the wild stories told about themselves and about their khukuris. After hearing their own stories, one Gurkhali observed: "They feared us, but all said and done, if we can win by reputation, no one wants to kill people."

"Khukuris of the Gurkhas of India is a sword knife with a blade convex along the back and expanding so that a large part of the cutting edge is also convex. It is used in hunting as well as in war, and armed with this weapon alone, the Gurkha will kill a tiger single handed", mentions *The War and the Chase*, London's Homaman Museum, 1929.

An Indian officer once wrote: "Gurkhas are fond of three things – Football for leisure, rum for pleasure and chopping heads for soldiering."

Memories from a Recce in an Italian battle field tour in 1991 by Major SJ Stannard 10th GR go: "We climbed up to 'Point 73' and there in a shade of the old house saw where Rifleman Ganja Bahadur Rai $2/10^{th}$ GR, attacked two Germans with his khukuri, killing them both. He was killed himself immediately after but the weapon of one of his victims is proudly displayed in 10th GR Guardroom, complete with slash marks all over the stock.

After the initial Gurkha assault, a badly injured German was found in a ditch. He had been chopped badly with a khukuri and was dying. Padre Don Armando carried him up the hill to the village. The soldier survived and now runs a BMW factory in Krefeld West Germany!'

It has been said that no wise man ever questioned Gurkha's challenge. In the course of World War I, a soldier of the 1/6th Gurkha Rifles on guard duty saw the huge bulk of a ship going silently without light one night along the Suez Canal. "Halt, who goes there?", shouted the Gurkha.

When he repeated the challenge and cried, "Halt, or I fire" a searchlight was switched on in the ship. When an incredible British captain saw a rifle pointed steadily at his battle-ship, he brought it to a stop. It took some time before a Gurkha Officer arrived and the sentry called out, "Pass, friend, all is well"

As the battleship glided on, British soldiers on board cheered the brave little Gurkha standing stiffly at attention with his kike slung behind his webbing belt.

The tough heavyweight Belgian actor Jean-Claude Van Damme was being tipped for a star role in a screen role in a screen version of 'Swan Lake'. Here, he is seen with his kike posed for screen.

On the day Germany surrendered the west of Fort Thatmyo, Burma on 8 June 1945, the ambulance drivers of the American field service wrote the following poem in honor of the Gurkhas. These ambulance drivers were Fred Kern, Rufus Mathewson, Duncan Mccollester, Robert T. Martin, III and Gene Ricca:

Here's to Johnny Gurkha

Here's to Johnny Gurkha
Who fights even to his grave.
Remember that Charge on Hill 565
The Kukris leaping as though alive.
Remember the marches under the Burma Sun,
And the battles that followed were always won.
The wounded with their grin and nod of their head

Forgetting the number of their own sates dead.
Thinking of their enemies' heads they did job,
Will answer you query, with Tekhai Sahib.
Gentlemen, stand raise your glasses
And drink this toast to those brave men
Drink to Johnny Gurkha, again and again

Shabaash, Johnny, Shabaash

There is another story of the Gurkhas of the WW II. It is said that the General Slim unveiled to his senior commanders, as far back as December 1944, his plans for his 'master stroke' thrust to Meiktila. Utmost secrecy had to be observed. It was not until the 1st Battalion was well on its way to the Irrawaddy at Pakkoku that the soldiers could be told about the plan. No doubt, Colonel Mc Cready told the British Gurkha Officers first, and they then briefed their men, thus:

In C Company that redoubtable soldier, Subedar, Rakam Sing Rai was company's 2 I/C. It is on record that he briefed his men thus;

'It is to be like boxing but without following the British Queensbury rules. You jab, jab and jab again to his nose, keeping his attention high.'

"Then you poke him in the eyes with your fingers so he cannot see. And then suddenly, you kick him in the balls. 'Ketaharu' or boys, our job is to kick the Japanese in the balls as hard as we can."

It is reported that the soldiers could not stop laughing at Rakam Sing's hilarious analogy. Every time the men of C Company saw Rakam Sing Rai thereafter, they stood to attention and saluted him with their right hand and covered the testicle areas with their left hands. This went on every day until the Irrawaddy was crossed and the operation started in earnest. I met Colonel Mc Cready DSO, OBE.CBE (CO 1/10th PMO GR 1953- 56) best friend of the late Maj. GCO Purne Rai WWII who received the Black Cat Banner for capturing the first alive Japanese soldier in the Burma War.

Letters from the Battle Field

During the WW I, the Gurkha soldiers were part of actions in various fronts in Europe, the Middle East and Africa. A Wartime Censor's Office was located at Boulogne, France to keep track of mail to and from troops, from the sub-continent in France and England. It was the responsibility of this office to seize letters containing secretive information about the war fronts and conditions back at home. The Censor Officers prepared frequent reports, which sometimes included lengthy extracts from letter, they read.

Letters written by Gurkha soldiers during the WW I provide the most direct written evidence found thus far for an examination of psychology, if you will, of Nepal's soldiers on the battlefield. The letters, which are excerpted below are from the stacks at the Indian Office Collection of the British Library in London. In the more than 20 volumes of censor's reports, each consisting of more than 200 folios, one can find only 50 letters from the Gurkha soldiers. The language in which the letters were written is identified as 'Devanagari', or Hindi or Nepali, and the extracts given below are in English translation.

Naturally, the Gurkhas thought a lot about the enemy. One put it this way, "On 9th day of May 1915, our Division was ordered to take German trenches at 5 a.m. Enemy trenches were 400 yards off from us. These trenches are near New Chaple. We went with fix bayonets and khukuris in mouths; this was the famous charge I have been through. We lost many men but we captured the enemy's line. I could not follow my company owing to sharp net bits struck on my right forearm, but it missed the bone, by mercy of Almighty God. Now I am in England and getting much better and shortly I will be back to France again and kill some more 'chutiya' … because they are not men because they use poisonous gas."

During the WW II, a 'sipahi' wrote to a friend – "I do not know if this letter will ever arrive because the censor may open it." A week later, he received a note saying, "The statement in your letter is not correct. We do not open letters."

Those who did not know how to read or write and their experiences are lost forever.

An Unknown Gurkha Warrior

Figure No. 23: *Statue of an unknown Gurkha soldier at the Horse Guards Avenue, nearby the Ministry of Defense, in the city of Westminster, London.*

There is another grave in the Gulf of Suez. A Gurkha lies buried since February 1915 on the outskirts of an Arab village. The death took place during a raid in the village. The Gurkhas had arrived by ship during the night; by noon the following day, the operation was over. Over 60 enemies were killed and 100 taken prisoners. While the Gurkha warrior was laid to rest, the Royal Marines provided a firing party and all British ships in the area lowered their flags to half-mast.

The regimental history made a note of the burial, "It is improbable that any Gurkha rifleman has ever been, or will be ever again, attended to his grave with so much honor."

During the WW I, Generals had been buried with less pomp and thousands of soldiers were hastily shoveled into ditches and covered up with little ceremony, noted a military historian. He added a rueful footnote, "unfortunately no one thought to record his name."

Lahore and Lahure

Gurkhalis from Nepal were first enlisted in the Army of king Ranjit Singh as infantry soldiers, as early as 1809. The word 'Lahurey' since then found its niche in the Nepalese vocabulary, as 'Lahore', now in Pakistan, was the capital seat of Ranjit Singh of Punjab then. There is an old song about the soldiers who returned home from Lahore on their leave. Back home for few days, they would bring new fashion and also love to their darlings waiting at home. Though the songs talk about Lohore and the persons who have been to this place as soldiers as Lahure, it also refers to the Gurkhas serving abroad in the past as well as now.

Lahureko relimai pheshan nai ramro
Rato rumal relimai Khukuri bhireko
Chyaat..........................[3]
Lahureko relimai pheshan nai ramro
Chhatre topi relimai Khukuri bhireko
Chyaat.........................[3]
Lahureko relimai pheshan nai ramro
Hatma banduk relimai Khukuri bhireko
Chyaat.........................[3]
Salko pata relimai tuppaima hariyo
Janda janda relimai dhawaima pariyo
Alu kati relimai tarkari tarera
Ayo Lahure relimai bairi lai marera
Chyaat.........................[3]

This melodious folk song cannot be translated exactly as it contains some typical mode of expressing love and wooing his beloved. However, it roughly translates: my darling look at me, a Lahure, (someone who has just returned from Lahore), how beautiful is my fashion, this red handkerchief and this khukuri tucked this my side, look how beautiful is my fashion, this round hat and the khukuri tucked into my side, look how beautiful is my fashion, a gun in my hand and khukuri tucked this side. Green is the leaf of sal tree on its top, Got the battlefield me in as soon as I landed, one makes a good curry of the slices of potato, here comes me, a Lahure defeating enemy.

When a Gurkha Fails to Return Home

Tens of thousands of Gurkhas never returned home and were consigned into oblivion forever. Their graves are scattered thick across the world. Whenever some survivors returned home, they had no words to explain. And this is how the late troubadour Jhalak Man Gandharva expressed his feelings to their concerned relatives in the song "When my mother asks". Jhalak Man Gandharva, a nationally known "gainay" or troubadour sang this folk song in Nepali through the Radio Nepal. His famous song can roughly be translated as thus:

"When my mother asks"

My mother will ask where her son is
Tell her that the battle has begun
My father will ask where his boy is
Tell him he's in the midst of the battle
Big brother will ask where his younger brother is
Tell him his inheritance has increased
Younger brother will ask where his elder brother is
Tell him he died in the battle
Elder sister will ask where her younger brother is
Tell her, she will miss the gift and me now onwards
Younger sister will ask where her elder brother is
Tell her that her maternal home has become smaller,
My son will ask for papa
Remind him to take off his cap
My daughter will ask for her father
Ask her to donate gold bangles
My love will ask for her husband
Tell her she is now free
Sister-in-law will ask where I am

She can kill a goat, make a party
My buddies will ask for 'Lahurey'
Tell them to kill all memories
Mother, don't cry, don't cry
If I live, I'll send you my photograph
Fathers cry for a year
Mothers weep their whole lives!

We, veterans and ex-Lahureys of the British Gurkhas in Kathmandu Valley collected and donated over NC Rs. 95,000 for Jhalakman's health care. Unfortunately, he expired in 2002; but his original and personal 'Sarangi' is bought by "The British Gurkha Study and Research Centre, Nepal" and is still in the safe custody.

The Gurkhas have become subject of songs, narratives, anecdotes and so on. Importantly, they have survived in the popular imagination of people. Books written about them by military officers have further multiplied and deepened the popular images of the Gurkhas.

Chapter VI

Modern Khukuribaajs

Officially the British Military had stopped its programmes, procedures and protocols of training one in the ancient khukuri combat techniques right after the WW I. It was natural that with the introduction of modern weapons, ages old technique and skills such as khukuri combat begun to die out.

However, for the Gorkhas, khukuri meant more than a weapon. This was the time when threat of the Imperial Japan to India was on a rise. At this moment, some seasoned Gurkha officers who deeply thought of restoring the ancient skills and tradition of khukuri combat came to play historically important role. They decided to re-implement khukuri combat during the World War II. Names of these seasoned Gurkha officers are:

1. Subedar (Capt.), later Major GCO, Rakam Singh Rai, OBI, IOM, MC, 1/10th GR. He was one of my uncles. His son WO 1 Manohar Rai (Retd.) lives in Kurseong Darjeeling, West Bengal, India these days.
2. Captain KGO, Kharka Bahadur Rai, MC 3/10th GR was my 'B' Company 2 IC in Majedee Barracks Johore Bahru, Malaysia once.
3. Capt. KGO Karna Bahadur Rai IOM, MC 1/10th GR whom I met with these KGOs/ GCOs during the Malaysia Emergency 1948-1958.

Implementation of khukuri combat technique was extremely vital to instill warriors' spirit in young Gurkha recruits who were mostly 15, 16 and 17 years old. This programme was fully supported by few young British Officers, namely; Lt. P. Dunkley MC 3/10th GR, later he became the pride of the 10th GR. This young BO had gained love and respect of the Gurkhas serving under him. They made him an honored member of the Rai and the Limbu tribes. Lt. J.F. Burns 3/10th GR and Lt. J. Wilson 1/10th GR., these young British officers fully participated every lecture and attended every phase of practice of live training with other Gurkhas.

With combat khukuri training, Gurkha officers instilled warriors' spirit of Rai tribes. Before and after every rigorous khukuri combat training, the phrase "Kaantar hunu bhanda marnu niko!" (Better to die than to be a coward) was chanted as the mantra. The glorious tradition of the Rais was indoctrinated and ingrained in all young Gurkha recruits during the War.

During the WW II in Burma Campaign, three of these under mentioned young British officers of the 10th GR namely PP Dunkley MC 3/10th GR, Lt JR Burns 3/10th GR and Lt. J. Wilson 1/10th GR had physically and mentally gone through the khukuri combat systems. They had become familiar with many rich cultural traditions, customs and rites of the Rais. Dr. M. Gyi recalls that Lt. P. Dunkley was awarded the MC medal for his outstanding leadership in securing the Fir Tree Hill. They were proud to see the Union Jack on the Hill. They saluted the flag and paid tribute to their fallen men. Lt. Dunkley became the pride of the 3/10th GR. This young British Officer had gained tremendous love and respect from the Gurkhas serving under him. They made him an honored member of the Rai and the Limbu tribes. 'Bir' Lt. Burns commander of the 'C' Company 3/10th GR was killed in an action at historical battle of Scraggy in Burma on 20 May 1944. He was only 22 years old.

Dr. M. Gyi (Maung Gyi)

Service No 105598 Rfm MANBAHADUR RAI 3/10th Gurkha Risles, Today he is known as Dr. M. GYI.

On this photograph, he is seen with all his personal collections of original 'kikes' from Nepal, India, and Buma. All these kikes' slender designs were favored by all Kiranti Rai tribes in the old Buma Gurkha Rifles. Here, he also displayed a symbolic arrangement of other 'kikes', representing a Gurkha patrol, along with two commandos daggers, representing a British officer and NCOs. Some of Dr. Gyi's 'kikes' are according to animistic belief, "spirited" with the souls of the enemy soldiers they killed.' Recalled By KKKG

Figure No. 24: Dr. M. Gyi (Maung Gyi) with his personal collections of kikes

Caption: Service No. 105598, Rfm. Man Bahadur Rai, 3/10 Gurkha Rifles. He is known as Dr. M. Gyi at present time. On this photograph, he is seen with all his personal collections of original kikes from Nepal, India and Buma. The old Buma Gurkha Rifles favored these kikes' slender designs. He displayed a symbolic arrangement of other kikes, representing a Gurkha patrol, along with two commandos daggers, representing a British officer and NCOs. Some of Dr. Gyi's kikes are according to animistic belief, "spirited" with souls of enemy soldiers they killed.

Dr. M. Gyi trained 10 kikes drill teams in ten different states in the USA. Each team had an average of six members. The teams consisted of young Americans from colleges, high schools and grade schools from Atlanta, Georgia. They enjoyed performing before the World War II veterans of the Burma campaign. The kike drill teams had also performed to raise funds for the Gurkha's Trust also. The very right hand man wears a 10th Gurkha Rifles (my Regimental) cap badge on his green beret cap. This photograph was received from my own respectable brother on 04 February 2003.

Figure No. 25: 10th Gurkha Rifle Badge (My personal badges on felt and beret hat)

Dr. M. Gyi joined the Army when he was 16 years. His grandfather and granduncle had also joined the 10th Burma Regiment of Madras Infantry in 1890. Man Bahadur Rai went through many intense combat khukuri trainings in jungle and mountain warfare at Shenagu Assam and training camp at Point 4241 on Yaigangpoki Saddle. Once, he was involved in a hand-to-hand combat against the Japanese fighting patrol, and was hospitalized with two broken ribs and dislocated shoulder.

At the time I am writing this section of the book, he is a retired professor at the Ohio University, USA. Whilst he was teaching at the University, he and another professor emeritus organized several Khukuri Drill Teams in 12 states of the USA. Originally, he had his own Khukuri Drill Team to achieve his three major aims:

1. To honor the spirit of Lt. Kul Bahadur Rai 1/10th GR, and Lt. Kulaman Rai 4/10th GR., both of his brothers who were killed in an action during the Burma Campaign War.
2. To honor the spirit of all brave Gurkha soldiers of the 10th GR, the brave British and the American soldiers who fought and died for their nation during the WW I & II.

3. To raise funds for the Gurkha Trust Funds. His teams had performed before Viscount John Slim, President of the Burma Star Association, Colonel William Dale, the former Commander of the 6th Gurkha Rifles and former director of the Gurkha Trust Funds in the USA.

His members of the drill team were made of young American civilians consisting of both male and female high school and college students, government workers, law enforcement officers, lawyers, businessmen, ex-servicemen and others. They studied history of various American and British Units that had served in the China-Burma-India Theatre of WW II. His Khukuri Drill Teams also performed at many annual re-unions:

1. US China-Burma-India (CBI) Veterans Association
2. US Merril's Marauders Veterans Association
3. US 1st Air Commandos Veterans Association
4. Burma Star Veterans Association
5. US OSS 101 Veterans Association
6. Regional and State CBI annual unions and other veteran's re-unions

Figure No. 26: Dr M. Gyi training with his green beret cap

The Americans and the British veterans of the Burma Campaign were very familiar with the name Gurkhas and their khukuri or kukri.

Dr. M. Gyi taught his students what he had learnt from his gurus during the Burma Campaign: Capt. Rakam Sing Rai OBI, IOM, MC 1/10th GR; Lt. Kharka Bahadur Rai MC 3/10th GR; Lt. Ratna Sher Rai MC 3/10th GR; Sgt. Karna Bahadur Rai 10 M, MC 1/10th GR.

Dr. M. Gyi was officially invited by the King Abdullah of Jordan to train his Special Forces about khukuri fighting skills. King Abdullah himself was his student. HM Abdullah and his younger brother Prince Ali learnt several skills and techniques for close quarter combat with khukuri from him.

Among several stories Dr. M. Gyi told me about the Gurkhas demonstrating their skills of kike in the War, I remember mainly three great heroes who had fought with kike and won the heroic status.

1. Rifleman Narpati Rai

In March 1944, during the battle on the Fir Hill Saddle against the entrenched Japanese positions, several small actions were taking place on the surrounding hill. One of the hills was called Right Hill, because it was on the right side of the Fir Hill Saddle. Two Japanese machine gun nests were strategically positioned between large boulders on the Right Hill to stop any flanking movements of the Gurkhas.

A section (squad) of young Gurkhas, from the 1st Platoon, B Company, 3/10th Gurkha Rifles was ordered to silence the enemy's machine guns. The Gurkhas gave covering fire as a fleet footed Rifleman Narpati Rai, only 16 years old, with his drawn khukuri, succeeded in moving behind the Japanese positions. He jumped between the boulders and took out one machinegun nest. Under covering fire, he then ran about 50 or more yards towards the other. As he attempted to take out the second machine gun position, one Japanese soldier stood up, instead of shooting at the Gurkha, he thrust him with a bayonet. The Gurkha rolled under the bayonet and cut the Japanese, but the other Japanese from the machinegun team finally got him.

Witnessing this, the young Gurkhas from the section, stood up and charged at the Japanese killing every one. This daring incident was reported to the Gurkha Lt. Chhabilal Rai. The report was never filed because the Lt. was killed two days later at the McMahon hill on Milestone 98.

2. Rifleman Chitarman Rai

After the battle of the Fir Hill Saddle, the Gurkhas from C Company were detailed to remove the dead and wounded, friends and foes alike. In one large Japanese bunker, which housed Japanese junior officers, they found the body of one young Gurkha Rifleman Chitarman Rai (17 years old) covered by seven dead Japanese. During the height of the battle, he crawled into the enemy bunker and killed the occupants with his khukuri before he was bayoneted to death. This incident was also reported to Lt. Chabilal Rai.

3. Rifleman Kitab Sing Rai

During March 1944, units from the Japanese 215 regiments had succeeded in setting roadblocks along Milestone-98 and Milestone-100, isolating the 37th Brigade. On March 24, B Company from 3/10 GR and C Company from 3/5 GR were ordered to clear the roadblocks.

The 2nd Platoon from B Company, 3/10 GR was sent out on a reconnaissance mission. As they reached the hills around Milestone-98, the Japanese ambushed them. Even though several Gurkhas got wounded, they managed to secure a small hill to set up a better defensive position. Soon they came under the fire from all sides. The Gurkhas fought back, using their ammunition sparingly. They radioed for reinforcement, but no help came. The Japanese forces had set up strong roadblocks stopping all reinforcements into the area. Night came quickly over these jungle - covered mountains of Burma and India. The Gurkhas repositioned themselves behind large trees for cover and established quick signal system to identify each other in the dark.

Soon, two Japanese "jitter" parties came up the hill to harass the Gurkhas. They shot flares, threw grenades at short range and ordered the Gurkhas to surrender. The Gurkhas remained silent and did not return fire for fear of revealing their position.

A young Rifleman Kitab Sing Rai (17 years old) decided to take on one Japanese "jitter" party, consisting of 3 men, who were only a few yards away from him. The Japanese were smoking cigarettes and talking. They did not realize that they were very close to the Gurkhas. Kitab Sing Rai silently crawled between several tall trees and thick bush and moved behind them. After a few minutes, the Japanese started yelling;

"Hey you Gukas, you surrender, you surrender".

"Hey Gukas, do not fight for English dogs. You surrender."

The young Gurkha quickly moved in and silenced them with his khukuri before they could react.

Morning came quickly. With tank support, men from 3/10 GR and 3/5 GR succeeded in breaking through the Japanese roadblocks and attacked the hills around Milestone-98 and 100. The 2nd Platoon was rescued.

The platoon leader reported the action of Rifleman Kitab Sing Rai to Capt. Jasna Bahadur Rai of B Company, 3/10 GR. It was never officially filed. The next day, both the Capt. and the Rifleman were killed on the same hill.

The true sons of Kirantis prefer to die a heroic death.

"The great men's experiences are small men's examples."

Living with Images

Images of khukuri and the wars Gurkhas participated have passed through various mediums. As a member of the family that has long history of serving the British Gurkha, I grew up listening to my brothers' experiences of living through the war.

My brother Man Bahadur Rai recalls the following episode, I put it in his own words: "One evening, our Battalion Commander Lt. Col. Roy Cousens soberly told us that the British Prisoners of the War in Singapore were being brutally treated by the Japanese. Thousands had died in slave labor camps from disease, starvation and torture. More than 200 Australians, British and Americans were beheaded, and many Gurkha prisoners of the War had also been shot, bayoneted and beheaded. I wept, I knew then I would never see my relatives again. Many men in our company remained silent with anger and tears.

The next evening, at midnight, our Commander Lt. Bakhat Bahadur Rai led us several miles along a jungle trail into the valley of caves, which was located to the northeast of Shenagu. When we reached there, we were surprised to see a large gathering of some Gurkha soldiers from other Battalions.

A Capt. from 3/5th Royal Gurkha Rifles led us in a sacred blood oath ceremony. This was done in a great secrecy, as the officers disapproved our traditional "pagani" or customs and rituals of evoking spirits. In a priestly voice, he evoked the Lord Veen or Shiva (the destroyer of evil) and 'Yama' (the king of death). We raised our khukuri, cut our fingers and smeared blood on the blade of our khukuri. We formed a long column. One by one, each man thrust his khukuri into a small flame placed before the Capt. He sprinkled the holy ashes on our heads.

Steel and spirit became one in this way; our souls got united with our khukuris. We made a solemn oath: "the shame of Singapore would be avenged; the Japanese would be defeated".

There are several other stories, Man Bahadur, my brother further recalls the episode associated with "the Burma Campaign during World War II" in the following manner:

"I was reassigned to the Platoon 1 under Sergeant Bakhat Bahadur Rai in "C" company. Our company was to secure the left side of the saddle, which was dubbed "Left Hill." In three columns, 1st, 2nd and 3rd platoons advanced halfway up the hill without opposition. Suddenly, the Japanese opened fire with automatic weapons from two bunkers near the top. Several men from the 2nd platoon fell. Then Sergeant Bakhat Bahadur Rai, leader of our Platoon, ordered us to flank the bunker in front of us. We crawled toward the summit".

"Three or four Japanese from an adjacent trench stood up and threw grenades. The machine gun from the bunker suddenly turned on us, wounding several men and pinning us down. Then, one fleet-footed

Gurkha, Narpati Rai, stood up with his khukuri and charged, yelling "Ayo Gorkhali!" meaning 'Here comes the Gurkhali' and jumped into the enemy trench and wiped out the Japanese machine gun team. Inspired by his daring example, we all fired our weapons and charged, while the 1st and the 2nd platoons rushed towards the other bunkers. Soon, the Left Hill was secured. The Japanese lost 32 men, most of whom we had cut down in their trenches and bunkers with our khukuris. Surviving Japanese ran down the hill into thick forest. Our sergeant was pleased with our first battle performance".

Man Bahadur Rai further recalls: "According to the 10th PMO GR's Roll of Honor 1890-1994, 73602 Rfm. Narpati Rai 3/10th GR 23.3.44 KIA was killed in that action in Burma".

He recalls another episode of the war he fought in the Gibraltar Hill:

"The Gibraltar Hill was one of many hills along the Tamu Road controlled by the Japanese. The Hill is about 6000 feet above the sea level with several high cliffs, dotted with huge boulders. There were three plateaus on the Hill. The Japanese gun positions were on the 3rd plateau. Most of the trees had been blown down and burned by previous bombing and shelling by the British planes and artillery. The Japanese bunkers and trenches were still intact with some negligible damage.

A company of Japanese troops occupied the hill. For several weeks, they had been firing on the British convoys, and had destroyed 3 trucks, 2 light tanks and 1 ambulance transport, killing some 20 Indian and Punjabi soldiers. They had also been shelling the British positions on adjacent hills. Several infantry assault by the men from other battalions had failed to take the Hill, resulting in high casualties.

The 3rd Battalion from the 10th Gurkha Rifles was ordered to try their turn. On early morning of May 24, 1944, three platoons from the C Company of the 3/10th Gurkha Rifles moved up the mountain. The 1st Platoon advanced from the south side of the Hill. The 3rd Platoon provided close support with ladders to scale the cliffs. The 2nd Platoon was put in reserve.

After 6 or 7 hours of hard climbing, the 1st Platoon reached at the second plateau. From concealed bunkers and trenches, the Japanese launched their mortars and grenades, followed by machine gun and rifle fire. Many Gurkhas fell during the first assault.

Most trees in the Hill had been blown down and burned by previous artillery shelling. The men had little cover. They tried to advance up the Hill but were pinned down by the automatic weapons fire.

Suddenly, it could be of divine intervention that the dark clouds and an early monsoon mist began to settle around the entire mountain. Under the cover of drifting mist, the 1st Platoon silently moved to the northern slope. The fire teams from the 2nd Platoon moved up behind large boulders, and started firing at the Japanese positions above. They intensified their fire giving cover for the 1st Platoon to move undetected.

The Japanese soldiers continued to fire and threw grenades down the Hill, thinking that the main assault would come from the south side. They did not expect the Gurkhas could climb from the northern slope, because it was too steep, slippery and treacherous. After several tries, the men from the 3rd Platoon were able to stabilize two long ladders along a steep cliff. The 1st Platoon moved up the ladders and reached the third plateau where the Japanese were still shooting and throwing grenades down the Hill.

The enemy bunkers and trenches were only a few yards away. The monsoon mist became thicker, making visibility very poor. The platoon leader, Sergeant Bakhat Bahadur Rai ordered the men to draw their khukuris and not to fire their guns.

Suddenly, the Gurkha Sergeant jumped up and yelled, "Gorkhali Ayo!!!" All the Gurkhas in the platoon yelled in unison and charged the bunkers and trenches from all directions. The surprised Japanese fixed their bayonets and drew their swords yelling "Banzia! Banzai! Banzai!". The Gurkhas only heard the sound that meant nothing to them.

With drawn khukuris, the Gurkhas charged slashing, chopping and thrusting at the confused enemy. The Sergeant shouted out "Ayo Gorkhali! No prisoners!" The Gurkhas followed, shouting "Ayo Gurkhali! No prisoners! No prisoners!"

There were violent collisions of steel against steel, steel against flesh, and flesh against flesh – it was a killing frenzy between the soldiers of Her Majesty and the warriors of the Rising Sun. Many Japanese escaped, running and rolling down the slopes. They left 145 of their wounded and dead behind. The Gibraltar Hill was finally taken after a hard fought and hand – to hand battle.

The commanding British Officers, Colonel Cosens and Major Edwardes congratulated the men of the 3/10th Gurkha Rifles and declared that their performance was one of the finest examples of courage in the regiment's history. It was also considered to be one of the most successful "Kukri Charges" against the well-entrenched enemy.

Later, Sergeant Bakhat Bahadur Rai, leader of the 1st Platoon, C Company was awarded with the Military Cross Medal for his outstanding leadership and courage in the battle.

General Slim's "Defensive-Offensive Strategy" was brilliantly implemented. The British Commanders, who served under him, performed their duties with unwavering purpose in defense of India. His well-trained and well-disciplined men fought with unequalled courage and tenacity against furious Japanese assaults.

The warriors from the Japanese Imperial 33rd Division could not secure their hold on the mountains around the Shenam Pass. On hill after hill, they fell to the men of the 23rd Division of the British XIV Army.

The battle for the Imphal region resembled medieval battles where close quarter hand-to-hand combat was the order of the day. The courage and skills of the Japanese jungle fighters were matched against the British trained Gurkha soldiers. The name 'Gurkha' and their khukuris and their battle cry "Aayo Gurkhali" instilled fear in the heart of every Japanese soldier who faced them in battle.

By the end of July 1944, the Imperial Japanese 15th Army was decimated. Thousands of their most seasoned soldiers died from wounds and disease. The remnants of the once victorious army retreated back into Burma. With the battle of Imphal, the Imperial Japanese Army suffered their greatest land defeat during the WWII.

The Battle of the Gibraltar Hill was one of many examples where the Gurkhas fought with loyalty, courage, skill and tenacity. This small battle, along with many other battles on the mountains of Shenam Pass will always be remembered as "Scraggy" in the annals of the 10th Gurkha Rifles" told Man Bahadur Rai to me.

Khukuris and Samurai Swords

The Japanese soldiers in Burma were courageous, determined, experienced and skillful in jungle fighting. They did not know the meaning of surrender. They considered themselves as the 'invincible warriors' of the Rising Sun. General Slim surmised that the over-jealous Japanese commanders, infected by 'victory disease', would push their soldiers to their physical and psychological limits. This Japanese code 'Bushido' inspired many of their victories against the British, French, Dutch and Americans in East Asia and Southeast Asia. The soldiers would eventually breakdown in the most formidable jungles of Burma and India.

The Kiranti warriors, on the other hand, believed in the spirit of their Veen or the God of destroyer, and the Samurai warriors in their spirit of 'Bushido' the Samurai's code. Both of the party did not fear with death, and they did not know the meaning of surrender. They believed that they were invincible and no other warrior in the world could defeat them in any battle. The Kiranti warriors believed in one word "the victory is always ours in any battle." But, the battles do not determine who is right, but only who is left. The westerners think that the Japanese and the Gorkhali were violent killers which is not so. The east or the west, there is "no gentleness in war."

The Gurkha soldiers, trained in both jungle and mountain warfare, and also honed in close-quarter combat with khukuri, played a major role in shattering the Japanese myth of 'invincibility', and their dream of conquering India.

I believe that both of our ancient swords no matter what shape or size, but all in all, their effectiveness is very equal, both in sharpness and in 'taste' both sweet and sour. Both do free ones from fear, grief, shame, and purify warriors' deeds. So much so that the British writer and war experts regarded the Gurkhas and the Japanese as sturdy simple dumbs.

"Untrained intelligence is as much of use as a bottle without an opener", wrote the Duke of Edinburgh in the Wisdom Magazine.

I also believe that Japan and Nepal were the only two countries among others who use their ancient weapon and gave unique demonstration to rest of the world during the WW II.

Japan is famous for her Samurai. Her principal weapon is the sword. Sword making in Japan has a very long and glorious history.

Among many Japanese fighting generals during the WW II, Lt. General Renya Mutaguchi was one of them. He fought in China, Malaysia, Singapore, and during the Burma Campaign against the Allied Forces. He acted on the order of General Tojo, later Premier and Japanese war minister. Mutaguchi was a man without fear who actually loved fighting entering a fray with exaltation. He was a big man with strong personality, forceful, self-confident, over bearing and also a bully. Once he was wounded with LMG bullet on shoulder but refused to be evacuated.

He always tried to lead his Division from the front. In the attack at Keppel Barracks, a staff officer tried to dissuade him from going forward, saying, "The shellfire is heavy, your subordinate commanders will think you have come forward to spur them on whereas they are doing all they can and can do no more." To this, Mutaguchi replied, "I do not wish to supervise the actions of my regimental commanders. But it is possible that they will both be killed and I wish to shake them by the hands before they die." Once he said, "I have killed thousands of my men; I should not go back to China alive."

Mutaguchi ended the war as commandant of the Junior Course at the Japanese Military Academy. For 18 years, he remained quiet about the war. 'Despite constant criticism of any actions', he said, 'I have preferred not to comment on the Campaign." In Japan, there is a saying "The defeated General does not talk of war." It is a good maxim. Mutaguchi died age 76 in August 1966.

Lt. General Shozo Kawabe, the Supreme Commander of the Burma Area Army had absolute confidence in the invincibility of his troops. He boasted that his forces would capture the Imphal region of northeastern

India within three weeks. He also declared that the conquest of India would be his birthday gift to the Japanese Emperor Hirohito.

In my view, Gurkhas fought with the Japanese warriors and vice versa. Many Japanese died. So did the Gurkhas in the war. Brave were the Japanese. Great is the heritage the Samurai sword has created. So were the Gurkhas and their khukuri. The Gurkhas and the Japanese could have become enemies for ages. But the reality is that both of the parties were dedicated warriors. They fought to prove their self. Any party could have become the winner. Gorkhas fighting for the British Empire respect the Japanese warriors. Their khukuris and the samurai swords of the Japanese are made of the same material. Both khukuris and samurai swords evoke the same spirit. As a retired British Gurkha I respect the Samurai sword.

My Malaysian Experience and More

As a British Gurkha, I also lived through the war similar to the one my brothers narrated to me. This was when I realised the real importance of khukuri and what possible help it could have provided to my brothers and other Gurkha fighters in the Wars.

When India obtained its independence on 15 August 1947, the British Government was keen to obtain the services of some Gurkha Regiments for their commitments in Malaysia, Hong Kong, the Far East and the United Kingdom. Negotiations were carried out between and among the representatives of the Government of India, Nepal and His Majesty's Government of the United Kingdom on future employment of the Gurkhas. The British War Office specified as early as 8th August 1947 that it preferred to have the 7th and the 10th Gurkha Rifles which recruited the Kirantis from the East and the 2nd and the 6th Gurkha Rifles which recruited the Magars and the Gurungs from the West. This enabled the British to spread as widely as possible over the recruiting, rang rooting areas in Nepal.

In June 1948, with dramatic speed, the future of Malaysia was threatened by the Communist insurrection. Within days, the security forces and the Malaysian Government were fighting hard to prevent

Ching Peng and the Communist bandits-soon to be called Communist Terrorists [CTs] – from overrunning the country. The Gurkha Infantry Battalions were so hard-pressed that even their recruits were sent out to guard vital points, and on some occasions, without ever having fired their weapons on a range. Not surprisingly, they were ordered to use their khukuris and bayonet in an emergency.

The Gurkha Training Centre in Sungai Patani (North Malaysia) was established on 14 August 1951. From then onwards, all rang roots recruits for the Britain's Brigade of Gurkhas were given 10 months' hard trainings before they joined their regiments. The monthly salary of the Gurkha serving in Malaysia was then Malayan Ringit (M$) 42 per month as against the M$ 450 received per month by the English soldiers.

In the early days of our 10 months' recruit's training period in Malaysia, the military issued khukuri had to be displayed on charpai or a coir bed. Khukuri had to be laid out on coir bed charpai, i.e. the sheath, *farash*, *karda*, *chakmak*, and khukuri for the section commander's daily inspection. The *farash*, sheath, pommels of *karda* and *chakmak* had to be highly polished. Boot polish, candle, water and soft clothes were the best materials to be used.

Khukuri would be carried by rang root only during the line or barrack sentry duties. It was also carried during the final Passing Out Parade. The Champion recruit of that particular year used to receive a special army issued khukuri through the hand of Major General of Brigade of Gurkhas.

For generations, the Gurkha regiments in the Indian Army had arranged their own khukuri supply through their depot's blacksmiths who used to provide blades of original and in traditional form in length and weight, which cut almost anything and was of deep concern to an enemy. But it was not the case in Malaysia. We still remember the Brigade's first massive delivery of ordnance issue khukuris in the year 1950 in Malaysia. They looked like khukuri, they felt like khukuri but they had one most important disadvantage. They did not cut either string or even any soft items. They were blunt and useless. A prolonged howl of complaints arose from every Gurkha regiments. The British Officers in the Gurkha regiments knew that Gurkhas were to have the best of everything and their khukuris are the best of all. I thought clearly something had gone seriously wrong between the intentions and production.

In peacetime, these issue or ordnance supplied khukuris are 'kalai' or electro-plated and used only for ceremonial parade and for the VIP guard duties.

Kike was used for clearing dense jungle for preparing Helicopters LZ, LP, and DZ to airdrop our re-supply of ammunitions, composite rations, medical aids, clothing and mails. I still remember the 'LS' (Long Surviving) Gurkha's rum, which is also one of our close and best friends in any hard time. I can't forget it.

Helicopters used for these operations were Scout, Gazzle, Lynx and Belvedere, so helicopter's marshaling procedure was also important. Without any further support of our silencer weapon, it was not possible for us to build rafts, wooden bridges across flooded streams, to chop woods for preparations of bunkers, sentry's posts, fence building, making jungle tracks, paths for scouting and patrolling.

Killing and slaughtering of living creatures found in the jungle, mountains, beach and sea was the only one means of surviving in the jungle of Malaysia and Borneo.

In Malaysia, during peace time, the Gurkha troops were supplied with live goats, pigs and our old favorite "Dumba", an Australian / New Zealander sheep for fresh meat. On every Wednesday and Saturday, these live animals had to be slaughtered by the fatigue party assigned for the Battalion's main cookhouse. Because of its thick wooly neck, sometimes our striker found difficult. However, there was no problem for those 'lathro' experts to deal with 8-10 sheep waiting in queue, and had to use sharpening stone often.

General public's eyes would always seem to be on the daily newspapers for the following kinds of news headlines during the Malaysian Emergency:

1. The bandit got a big shock
2. Bandits shudder at very mention of 'Gurkha'
3. Gurkhas kill 6, capture 3 in one battle

4. Gurkhas bag their 201st to top score
5. No. 12 of Manap's gang is killed
6. M$ 10,000 Bandit leader is killed
7. Rifleman Bharna Bahadur Rai of $1/10^{th}$ GR killed
8. Manap Jepun ('The Jap'), a notorious communist terrorist in Malaysia, who was said to have been immune to all but bullets of gold or silver. The reward was offered M$ 75,000.

Such was the public response to the Gurkhas. The Gurkha had the highest standard of jungle warfare tactics, skills and marksmanship. They knew how to live, fight and survive in the Malaysian and Borneo jungles. Secondly, the brilliant leadership of the seniors as well as courage, discipline and devotion of soldiers proved that the Gurkhas were again the best in the whole of British Army in dealing with the CTs in Malaysia.

The result was welcoming. The CTs were driven further and further into the interior, and ultimately they became a small group of extremist of CTs in 1958, and escaped over the border into the jungle of Thailand. Peace was restored in Malaysia. Today, both the young British National Servicemen and the Gurkhas can look back on their service in Malaysia with justifiable pride. The days of confrontation, of blood, of sweat, of tears are in still in their memories.

During the ten years of emergency and confrontation, many Gurkha comrades gave their lives to preserve their Gurkha pride and regimental honor. "The terrors of the jungle today have largely gone. We know how to live and move in it. The jungle is simply another element in which men fight like in the sea, the air or the desert. The enemy may fill any of these elements with his hazards, but of the four, the jungle itself contains far less natural hazards than the sea, the air or the desert", wrote brigadier MCA Henniker CB, DSO. MC.

Even among many celebrated units of Britain's Brigade of Gurkhas, the 10th PMO Gurkha Rifles has an outstanding reputation as a fighting regiment, living up to the saying, 'the regiment is for service and not for show'.

When the last British Army band was recalled from Singapore, it was proposed that the Gurkha band represented Britain at an international fair. "Can you imagine that?" asked the Ambassador who vetoed the plan. We know the Gurkhas do all fighting for us now, but at least there should be a few Englishmen left to play "God Save the Queen" (*The Guardian*, June 14, 1974).

In Malaysia, during 1957, I was given a task to perform a khukuri dance / khukuri drills for the 'Kala Ratri' or the Nineth Day of Dashami. I was then the Army Number 21142158 Corporal Kaman Sing Rai, but I must say I was not as smart as a King's Corporal of WW I or a Queen's Corporal of the African War nor "Little Corporal" of the WW I.

However, remembering their examples, I organized a khukuri drill team. My aim then was to give our best performance for the first time to VIPs, Regimental guests, and to all families and soldiers of the 10th PMO Gurkha Rifles.

On the 'D' Day, our team marched on the stage in two files, me leading the team. As I remember, red handkerchief was tied on our both wrists and a shield on our left hand and shirupate khukuri on our right hand, the real 'phaudari' men's 'kaida' or procedure in real action, we started off our khukuri's actions with our old war cry "Aayo Gurkhali!"

Phase I

We demonstrated how to attack and block using both khukuri and shield, and also demonstrated few quick chops, thrust and slash on stationary targets.

Phase II

We then demonstrated about how to change into quick box formation to face fresh targets and then to continue the previous actions and so forth. The time allowed for my team was just for 10 minutes. We completed my mission at midnight when the lights were switched off for one minute to observe the actual 'Kala Ratri', the darkest night of the year that falls in the ninth day of Dashain.

Today, I remember Lt-Col. Reg Twelve Tree MC was then our Commanding Officer. He was the one who introduced Inter-company Boxing Championship for the first time in the 10th PMO GR.

During those years, I was also a boxer myself, but not as professional as late Lt. (KGO) Mek Bahadur Rai 1/7th GR, someone popular as well as a professional boxer in the whole of Far East Land Forces. My OC then was Major HCS Gregory OBE at one time, the great friend of late Major GCO Purne Rai of the 10th GR during the World War II.

Hong Kong

Our regiment, the 10th PMO GR was stationed in the Queen's Hill Camp (NT), Hong Kong 1967. On 24 August 1967, Lt. (QGO) Dhan Raj Rai, $1/10^{th}$ PMO GR, was commanding a platoon. His platoon was tasked to maintain a barbed wire obstacle erected on from the British end of the Man Kam to Road Bridge on the Hong Kong border.

Some forty to sixty Chinese workers attempted to remove the obstacles erected under cover of a barrage of stones. They fired a number of automatic weapons over the head of our troops from positions within 200 yards of the frontier.

Lt. (QGO) Dhan Raj Rai positioned himself some fifteen yards from the end of the bridge. Men of his platoon supplied the NO 80 WP (white phosphorous smoke grenade) and teargas grenades. He personally defied all initial efforts by rioters to dismantle the obstacles. Finally, they could not to stand up to the accurate barrage of grenades. They brought up trolleys of straws to provide covers and act as battering rams.

Lt. (QGO) Dhan Raj Rai attacked the rioters with WP grenades and forced them to withdraw. The rioters succeeded in cutting and removing a portion of obstacle and eventually resorted to attempt to apply a grappling hook and dragged the wire away. On three separate occasions, Lt. (QGO) Dhan Raj Rai went forward to the obstacles, grenade in one hand and khukuri on the other. He succeeded cutting the grappling ropes. Finally, despite all our efforts, the last strands of wire were removed. With Lt. (QGO) Dhan Raj Rai and his men guarding open at the end of the bridge, no one dared to advance on the British territory.

The personal courage, leadership and determination of this Queen's Gurkha Officer were the prime factors to nullify a violent and aggressive mob in an action that lasted nearly for four hours. Later, Major B. M. Niven MA was awarded the MBE medal, and Lt. (QGO) Dhan Raj Rai was awarded the Queen's Commendation.

During this particular action, I was then a CSM of the same company. At one stage, the bullets were damaging some pine trees on top of our command post. The one-sided firing went on and off for about two hours. All Gurkhas were on stand position and waiting for further order from the OC. We had permission now to return the enemy's fire.

Later, questions were asked to the Section and the Platoon Commanders: "Why they had not fired back?"

"Their bullets were 'Not Effective'; none of us got hit", was the reply.

To achieve peace in Hong Kong, the British and the Britain's Brigade of Gurkhas truly and respectfully guarded the British territories and its 235 other small islands for 99 years. Hong Kong was handed over to China by the British on June 30, 1997.

My regiment 10th PMO GR was stationed at Queen Elizabeth Barracks Church Crookham UK from March 1983 to April 1985.

Cyprus

In 1974, 10th PMO Gurkha Rifles was stationed in UK and was mainly involved in guarding Buckingham Palace, St. Jame's Palace, Tower of London with Windsor Castle and other public duties in London. At the same time, the Regiment was warned to move to Cyprus within shortest notice. The Regiment was mentally prepared for the task ahead, and Major M. G. Allen, who then was my OC, briefed the Regiment on the situation before moving on the ground thus: the troubled history of the island, and the events leading up to the Greek Coup and the Turkish involvement with Cyprus, the East Sovereign Base Areas (ESBAS), the Turkish Cypriot Freedom Fighters (TFF), the Greek National Guards (GNG) and the United Nations Forces in Cyprus (UNFICYP).

At the end of his briefing, a junior rifleman asked, "But who is my enemy?"

"This was a question which we never really solved to his satisfaction!"

Saturday 10 August 1974 was the Krishna Janmastami day (a Gurkha holiday).

But, "Go" came out of the blue at 17:30 hours.

By the early hours of Monday morning, we were all in the air on our way to the Mediterranean.

On the second day of arrival on the ground, the B Company Capt. QGO OC Major M G Allen (later CO 10th GR) and 2 I/C Chandra Bahadur Rai (later MVO Hon. Major GCO) were tasked in establishing vehicles checkpoints, and controlling refugees on the main road between Larnaka and Famagusta which runs 15 miles through ESBAS. On the day two, my No 6 Platoon received an order to prepare a hasty defensive position on a small hilltop, which was 300 yards away from the CHQ location. There were few thin pine trees around this hilltop, but the order was not to damage or use any khukuri. "As necessity knows no law", few pine trees were chopped for the GPMG post for the gunners to mark arcs of fire on the ground. Our Company's area was too large to defend north and west to meet the Turkish Army or south and east where the GNG were facing the TFF and 13,000 Turkish Cypriots besieged in Famagusta. As we were busy and at midday, the Turkish aircraft bombed and rocketed the GNG positions just three miles to our front. The effect on our soldiers was stimulating; those who had not completed their trenches did so without any further order.

Other companies were also involved in similar tasks in various other areas too. Alfa Company was tasked to prepare a campsite for 20,000 refugees at Athna Forest. During this particular period, combat khukuris were used for no other than under mentioned reasons. Khukuri was used first to clear off the undergrowth in and around the forest areas; and to safeguard all those large number of refugees; and for other security reasons.

By the way, no fresh meat was supplied to the Gurkha troops as it were done in those old days in Malaysia. However, the 'karda' itself played an important role as for corkscrew to open up those famous Cyprus red and white wine bottles' tops and to butter bread and biscuits

in mess hall. After 6 months of operational period in Cyprus, the 1st Battalion, the Devonshire and Dorset Regiment relieved our Regiment.

On arrival of 1st week at our Old Barrack at Church Crookham Aldershot, we found ourselves again rehearsing for changing the Guard duties with one of the English Regiment at Buckingham Palace. The RSM and CSM's ear shattering roars for ATTEN-SHUN! PREE-SENT-ARMS!! could be heard miles away from the Drill Square. In February 1975, the regiment completed UKLF tour.

Falklands Campaign

On 12 May 1982, the 1/7th DEO Gurkha Rifles left Church Crookham bound for Southampton and at midday embarked on the Queen Elizabeth II. As part of 5th Brigade, the Battalion set sail for the Falklands, which had been invaded and occupied by the Argentineans some five weeks earlier. The Gurkhas arrived at the San Georgia and they were transshipped from the Luxury liner to the San Carlos waters.

After landing at the San Carlos, the Battalion took over responsibility for the Darwin and Goose Green area, which had been taken with such skill and courage by the 2nd Battalion of the Parachute Regiment. In a series of spirited patrol encounters, the Gurkhas captured eight prisoners of war. Shortly afterwards, they were deployed to the east, moving partly by forced march over truly inhospitable terrain in appalling weather to take part in the final assault on port Stanley.

Mount William was to be their objective. It was a hill that was reported to have been occupied by a battalion of Argentineans. However, to the deep disappointment of the Gurkhas, the defenders decided to flee rather than face the enemy whose reputation filled them with awe. One of the last strong points guarding the port, Stanley was thus taken without a fight. The battalion's casualties were light, with one QGO and eight Gurkha from other ranks being wounded.

The Gurkhas played an important part, and their very presence helped to bring the campaign to a quick end, without unnecessary casualties on both sides.

In 1982, the young soldiers in the 1/7th Gurkha Rifles lived up to their motto:

'Kafar hune bhanda mornu ramro" or It is better to die than be called a coward.

The encounter in the Goose Green area itself was a shameless. There were 471 disgruntled Argentinean prisoners locked up in sheep-shearing sheds who were very frightened that their new guards were Gurkhas. Having read the magazine that featured an article headlined 'Los Baba'ros Gurkhas'. It described in detail that Gurkha took no prisoners. These dispirited men were flown out to the San Carlos and the 'Barbarous Gurkhas' could then get on with the job at hand.

A patrol from Reconnaissance Platoon ran into an Argentinean group armed with anti-aircraft missiles. The Gurkhas forced them to surrender by simply drawing a single khukuri and brandishing it aggressively. Later, that khukuri was to be auctioned to raise money for charity and fetched £ 1,200!

However, to cut the story short; afterwards, they (1/7th DEO Gurkha Rifles) were deployed to the east, moving partly by forced march over truly inhospitable terrain in appalling weather to take part in the final assault on Port Stanley.

Combat khukuri though has a cultural root going to the *Phaudari*, the Rais and the Limbus celebrated in the past. The British Gurkhas have made this tradition a part of the modern war technique. Fighting with khukuri has become a separate domain of study, training and innovation. One should thank the genius at the barrack and training centers as well as those in the battlefield for recognizing and proving the hidden spirit of the khukuri or kikes. They are the real khukuribaaj of modern times.

Chapter VII

Khukuris and Gurkha Identity

The force that makes Nepali khukuri powerfully effective emanates from its unique design developed over the ages. A broad forward weighed blade with a downward curving design, permits its edge strike always to target with an angle creating a powerful slicing effect. The wrist, thumb and fingers further enhance the striking force of this side-shaped blade. The chop is commenced with an open hand and the black back, paralleled to the forearm. As arms move forward, the minor digits close sharply, accelerating blade's speed. Combined all these, it produces a devastating blow with minimal efforts.

It is not the weight and the khukuri that alone makes it terrible at a close quarter. Much depends on the skilled techniques one uses to strike. Together, it creates almost impossible situation for opponents to parry.

The real taste of this blade is not like that of sweet kiss the mantis offer to its prey nor is it like the sting of a black widow. It is said that the blade's clean chop is very much like good cooking; it can be tasted, but not explained.

The Gurkhas know how to use khukuris during the battle, and bring victory in their favor. They also know how to exhibit the art and technique of using it. They honor their kikes at home, on social and culturally important ritualistic occasions. It is time that the authority concerned realised its significance in the global context. One should explore the heritage of khukuri making arts and skills as well as the history carved out by the Gurkhas in order to enlarge Khukuris' market, scope and so forth. Moreover, khukuri or kike needs to be taken as heterogenous weapon. There is should be official institution to recognize, categories and promote the most authentic khukuri. This has become necessary to reward the artists who insert their skills and knowledge in their creation. It demands upon the authority to create a line between the artistic and genuine khukuri from the ones produced in mass in factory. The blacksmiths who have nurtured this skills and wisdom of creating khukuris or kikes need to be recognized and treated respectfully. Their kikes or khukuris should be promoted. Moreover, we live at times when people love and are ready to keep genuinely produced artifacts. Khukuri or kike has already become artifacts. It has already become a globally recognized weapon. Nepal needs to tap this opportunity, and bring prosperity to the people mainly the blacksmiths who have remained marginalized for ages.

One thing we are sure is that the world always looks for quality manpower. So does the British Gurkha. It is sure to embrace cost effective professionalism in the days to come. Warriors from Nepal are sure to meet the expectation of the British Army. I believe that there will always be a space for the Gurkhas. New generations Gurkha soldiers will keep on serving with moral high and pride. It is true that we live at time when there is not much war going. This means not many soldiers are required. In such context, only the cream of the crop will get selected in the British Force.

At the end, as a retired British Gurkha, a Rai from the Kirant community, and importantly someone who has vowed to honor kike with love and pride, I would like to tell my Gurkha boys and the girls that as aspiring British Gurkha and Indian Gurkha for that matter, you should own up the history. It is your turn to take your kike with love and pride. With this, you will salute and honor their ancestors. Pleased will be our ancestors. With this, you will honor the soul of the warriors who are buried all over the world over the centuries. With this you will evoke their bravery and honesty. With this, you will bring peace and prosperity to this world, and also to their home. And, with this you will please the lord Veen.

Glossary and abbreviations

Aptungsang	Rifles, guns, in Kiranti Rai's Bantawa language
Adda	Territories
Ayo Gurkhali!	Here come the Gurkhali
Ankhole	Khukhuri made in Ankhola village
Ashala	Fish (Khukuri shaped like a 'ashala' fish)
Achar	Chutney
Aran	Khukuri manufacturing factory
Bairi	Enemy, foes
Bhojpure	From the Bhojpur district (khukuri/ people)
Buduna	Bellied fish-shaped khukuri
Berha	Handle, pommel
Betho	Khukuri in Thulung Rai's language
Baaj	An expert, an eagle
Bhokshi	Chakmak in Bantawa Rai language
Biswokarma	Blacksmiths
Boju	Grandmothers
Baje	Grandfathers
Basha	Resting/sleeping place prepared from wood and poncho cape
Bhat	Cooked-rice

Bozo	Corus (alamus used for medicine)
Bulaki	Gold ornament women wear on their nose as a religious mark
Babi	Pig in Malaysia word/language
Bahadur	Brave, it is not an individual name and does not belong to any known religion, custom or society. He is simply from the hills and sometimes called 'Bir' too.
Bhutte	So blunt
Binayo	Bamboo's Harp (made from 'malingo' or bamboos)
Birkhe topi	Kilmanuk, (Pillbox cap)
Blood sport	Those involving in the killing of animals
Bhook or dhariya	Fish trapping baskets made of bamboos
Baba or babu	Father
Boka	Male (he goat)
Chakmak	Dull-edge thick knife (Hindi speaking people and Turkish call it chaqmaq)
Chainpure	From Chainpur in Sankhuwasabha district (Khukuri / people)
Chitlange	From Chitlang village (Khukuri / people)
Chub	Khukuri in Sunuwar language
Chimti	A twizer
Chiraito	Swertia, medicinal plant
CO	Commanding Officer
CTs	Communist Terrorists
CSM	Company Sergeant Major
CHQ	Company Headquarters
Capt.	Captain
C in C	Commander in Chief

Cummerbund	Waist rapping clothe
Cho	Notch in Tibeto- Burman (non-Nepali word)
Choli	Clothing that mother, brother gifts to daughters or sisters
Dhaakar	Heavy load carrying basket carried by porters in Nepal
Dhankutte	From Dhankuta district (Khukuri/people)
Dhal	Shield
Dhaago	Thread
Daabe	Khukuri in Bantawa Rai language
Dukuku	Karda in Bantawa Rai language
Dalchiin Dhungo(a)	Flint
Daap	Sheath, scabbard, case
Dasain/ Dashera	A Hindu festival
Danra	Hill
Dumba	Sheep (thick headed person) (slang)
DZ	Dropping Zone
Dharso	Line
Dhaami	Medicine man, priest
DSO	Distinguished Service Order (medal)
Dandi biyo	Game of stick, also recognized as the national game of Nepal
Dhawa	Battle, war
Daan	Ritual offering
DFH	David F. Harding, MA FRAS (Captain/Retired, 1/10 GR)

Elam	In the earlier days, Ilam used to be called as Elam, Phakphok or Charkhola district of east Nepal, (Old fort at Shanguri Bhanjyang)
Fish plate	The flat piece of iron that connects railway tracks
FM	Field Marshal
Ganjewal	Sherpa's style of Khukuri (introduced by Mr. Ganje Sherpa of Solukhumbu in the earlier days)
Gooja	Khukuri in Tamang language
Gothale	Cowboy/herdsman
Gundruk	Dried vegetable leaf
GM	Gurkha Major
GR	Gurkha Rifles
GPMG	General Purpose Machine Gun
Gallawala	Recruiter
HQ	Headquarter(s)
IS	Internal Security
IOM	Indian Order of Merit (medal)
IDSM	Indian Distinguished Service Medal
Jhankri	Witch doctor or faith healer, also called priest by Kiranti Rais and Limbus
Jhulo	Fire tender
Jaarh	Millet/ rice wine
Jahat	Trouble maker in Malayalam
Juwa	Gambling
Kaura/ri	Notch
Kike	Ancient Kirant Rai's word for Khukuri

Kukri	Khukuri spelled in English
Khurmi	A thick curved sickle carried by every Kiranti woman or often called khurumbi or kombi too
Kalahi	Electroplated
Karda	A small sharp knife
Kalam	A pen (usually made from malingo bamboo)
Katle	Fish (Khukuri shaped like katle fish
Khurcheyo	Khukuri in Magar's word/ language
Khisha	Leather pouch to keep 'jhulo' and flint
Kochaa	Khukuri in Gurung's word/ language
Kothi	Decorative dots (or kothi mohora/ Muhunda khukuri)
Kaida	Customary, procedure
Kafar/kathar	Coward
Kalaa	Art
Kokro	Cradle/ pram
Kukur daine	Smilaceae (an edible plant) found in Malaysia and Borneo's jungles *hot khote* a tall thorny tree's shoot on the top is an excellent vegetable found in Malaysia and Borneo's jungles
Kambin	Goat in Bhasa Malaysia
Kapala	Head in Bhasa Malaysia
Kampong	Village in Bhasa Malaysia
Kirawa	Tiger (slang) in Gurkhali language
Kuire	English man / blue eyed person (slang)
Khoppi	Game of coins
Kusume rumal	Silk multi coloured handkerchief
Kami	Metal smith or Lohar of the Kami clan
KGO	King's Gurkha Officer
Khaarka	Grazing field
Khadgaa	A two-edged sword

Kattari	A triangular two-edged knife a sword with a blade of varying length and width sharpened on inside edge. Referred to in old text as a kora or khura
Ketu	Last 'graha', a heavenly body that affects our luck
Kinama	Boiled and dried 'bhatamas' peas
Kaura/ri	Small sea-shells, pronounced as cowry/cowrie
Kami	Black Smith (dangra) (slang)
K-rations	Field rations of concentrated food
Lathro	Devastating, lightning, clean chop or stoke with khukuri
Lahure(y)	Soldier
Lohars	Iron Smith
Lalang	Grass in the Bhasa Malaysia
Lt.	Lieutenant
LP	Landing Point, where helicoptor lands
LZ	Landing Zone
LMG	Light Machine Gun
Line boy	Sons, brothers, relatives of Gurkha officers, NCOs and soldiers, who are serving or had served with the regiment
Murali	Flute made from bamboos
Murchunga	Jews harp made from iron
Mahila	Lady/woman
Mooth	Pommel, hilt, handle
Madal	Gurkha's traditional two-ended drum
Maruni	Man disguised as girl/woman
Mundhum	Means the power of great strength as a truly, holy and a powerful scripture for Kiranti Rais and Limbus. The Mangpas, Dhamis, Jhankris

	Phedangbas equal to priest observed this ancient culture since ancient ages
Mantra	Vedic hymn
Mukhtiyar	C in C or Commander in Chief of the Army
Mem saheba	Gurkha's ways of addressing and respecting any British officers' wife
MC	Military Cross (medal)
MBE	Member of the Most Excellent Order of the British Empire (medal)
MID	Mention in Dispatches (the oldest British award for the service in action)
Ningure/o	An edible fern
NCO	Non Commissioned Officer
NT	New Territories
Nepal Chautho Tara	Star of Nepal Class IV (medal)
Orang	Man/ people in the Bhasa Malaysia
OBI	Order of the British India (medal)
OBE	Officer of the Most Excellent Order of the British Empire (medal)
OG	Olive green (colour of tropical uniform)
OC	Officer Commanding
OP	Observation Post
Phaudari	Close - quarter combat by using weapons especially with 'kike'
Paltan	Battalion, regiment
Patuka	A long piece of white / other colour of cloth wrapped around waist by both man and woman
Pheta	White or any other colour of turban
Poncho cape	Waterproof cape of light weight with slit for arms

Potong	To cut or slaughter in the Bhasa Malaysia
POW	Prisoners of War
PTI	Physical Training Instructor
QGO	Queen's Gurkha Officer
RSM	Regimental Sergeant Major (the army's most senior NCO)
Roal	Made from highly seasoned timber which was especially used during 'phaudari' action by the ancient Kirantis of eastern Nepal
Roman nettle	Rarely found in Nepal with female flower on head (The common ones are the 'patle', ghariya, bhangre)
Roman snail	Edible snail found during 'duwali' netting systems
Rote	Orderly, helper (sahayogi)
Rahu	The 8th graha or sub plannet. The demon that swallows the sun, moon and causes the eclipse
RAF	Royal Air Force
RAOC	Royal Army Ordnance Corps of the British Army khukuri designed as leaf of siru grass (the famous Khukuri used by Majh and Pallo Kirantis, most popular Khukuri of the Kirantis)
Shishnu	Nettle
Sahib/sahib	Gurkhas 'kaida' for addressing and respecting any officer
Sinki	Dried vegetables
Shokpa	Yeti
Shamma	Priest in Rai language
SAS	Special Air Service

SOP	Standing Operationl Procedure (guide book for poperational procedure)
SLR	Self-Loading Rifle
Sgt.	Seargeant
Sardar	Chief, headman
Shangri -la	Roof of the world
Tuppi	Pigtail
Trisul	Trident, a three pronged spear
Titepati	Asteraceae (used for jhulo) wooden, bamboos or silver can for drinking jarh (wine)
Tamma chhinai	To chop like bamboo's shoot
Thaili	A round type of cloth purse carried by Nepali women folks
Tarawar	A long thin sword (talwar)
Sarangi	Nepali musical instrument made from wood and animal-hide
Sagarmatha	Mount Everest, the world's highest peak 29,028 feet (8848 metre high) bears the name of George Everest, an English man and surveyor- general for India. The Tibetans and Sherpas call this mountain peak "Chomolungma" that means the mother of the world, winds and snows.
Second I/C	Second in Command
Veen	Lord of the Kiranti Rais
WW	World War I, II
WOI	Warrant Officer Class 1 (one)
WO2	Warrant Officer Class 2 (two)

Yakthumba	After they lost their power of administration, all Yakhathumbas began to call themselves Limbus.
Z (Zulu)	Old African war warriors
Zoro	Leather whip used by western cowboys

References

Wilbrahim, EGERTON. *An Illustrated Handbook of Indian Arms and Those of Nepal, Burma, Thailand and Malaya*. Bangkok: White Orchid Press, 1981.

B.R. Mullalay. *Bugle and Kukri: The story of the 10th Princess Mary's Own Gurkha Rifles*. London: William Blackwood and Sons, 1957.

Chappel, Mike. *The Gurkhas*. UK: Osprey Publishing, 1993.

Smith, David Eric and Brian Horrocks. *Britain's Brigade of Gurkhas*. London : Secker & Warburg, 1973.

Kern, Fred, Rufus Mathewson, Duncan Mccollester, Robert T. Martin, III and Gene Ricca 'Here's to Johnny Gurkha'. https://www.thekhukurihouse.com

Kandangwa, Ram Narayan. *The Kathmandu Post*, May 3, 2010.

Lawrence, J.M. "Lewis Millett; awarded Medal of Honor after bayonet charge". The Boston Globe. November 19, 2009.

London, Perceval. *Nepal Vol. II*. New Delhi: Asian Educational Services, 1993.

Mullalay, B.R. *Bugle and Kukri: The story of the 10th Princess Mary's Own Gurkha Rifles*. Great Britain: William Blackwood and Sons, 1957.

Parker, John. "The Gurkhas Fighting in Faraway Places".

Personal Interview with Man Bahadur Rai

The Rev. Wood. *Travels in India and Nepal*, 1896.

Slim, William. *Defeat into Victory.* London: Cassell and Company, 1950.

Soldier, the magazine of the British Army. 9 February 1987.

Thapa, Netra B. *A Short History of Nepal.* Kathmandu, 1990.

The 'Parbate', a weekly magazine

The Amrita Bazar Patrika, 9 November 1914.

The Bugle and Khukuri. The Journal of the 10th PMO GR 2002.

The Guardian, Friday, June 14, 1974.

The Sunday Mirror, 08 July 2007.

The Sunday Mirror, July 8, 2007.

The Times of 27 June 1857

This England, Golden Jubilee

Editor's Notes

The writer of this book comes from a family that has served the British Army for generations. Memebers of his family and relatives have experienced the war of one or the other kinds of global order for generations. He has grown up with thick descriptions and stories of war. Though he started serving the British Army from 1950s and got retired in the 1980s, he personally knew many retired British Gurkhas who held memories of fighting in the First and the Second World War. This means to say that he holds personal as well as collective experiences of war that this world so much talks about. He owns up the history of serving the British Gurkha. As an editor, I regard it as the backbone of this book.

More than that, he is a Rai by birth. His community celebrates kike as a sacred weapon. No ceremony is completed without it. As a Rai, he grew up knowing and practicing kike in daily life. As a boy he grew up watching *phaudari*, a khukuri-combat that would take place regularly in the Kirant region in the past. Now as an enlightened man, who has concentrated on kike for seven decades, he seems to discover that the very cultural fabric itself is the force behind enabling Kirantis, the Rais and the Limbus to become brave warriors all over the world now. He finds a deeper connectivity between khukuri or kike and the Kirantis. He thinks that one can't be understood without the other.

On top of that he is determined to write a book on kike or khukuri. He wants to pass knowledge and tradition of using kike to the next generation. He invents this role out of his own realisation. He has spent decades collecting materials for that. His visits to the museums of

London and Kathmandu and the Kirant region of Nepal have helped him find out historically authentic materials. He has observed historical items and drawn energy to associate him with the project he has vowed to complete. He has read articles and books written by renowned historians on the British Gurkhas. He has honestly mentioned the sources he has taken the lines from though some of the materials he cites don't contain publication details. As an editor, I would like to request readers to look at the honesty of this writer. And, I hope that this book inspires researchers and academics to catalogue and bring out important documents to highlight the role kike and Gurkhas have played in bringing the current political and military order in this world.

The sense of humor the writer creates while narrating incidents and objects, and the point of view he develops to convince his readers makes one realise that KKKGji holds a writerly self. I enjoyed working on his manuscript.

As far as my reading goes Mr. Rai has used several materails from the *The Kukri*, the Journal of the Bde. of Gurkha, and he shows his deep gratitude to the writers he has quoted and cited information from. He has consulted 'The British Gurkhas from Treaty to the Supreme Court' written by Shree Gyan Raj Rai (Rtd. WO II Ex- QGE). This important book was published in 2059 (Bikram Sambat) by the United British Gurkhas Ex-Servicemen's Association Nepal.

I thank him for trusting me to go through his manuscript and rearrange materials so that it comes out as a book. I thank Dr. Urvashi Rai, the US-based scientist, and importantly the daughter of KKKG for helping me to complete the work on time. I also thank Bidur Dongol, the proprietor of the Vajra Books, a Kathmandu-based publication house for connecting me with the world of Gurkhas and their khukuri, a rare moment itself.

Shiva Rijal, PhD
Lecturer
Central Department of English
Tribhuvan University

Many firsts

One cannot be many first like Charles, the son of HRH Queen Elizabeth II of the United Kingdom. He is the only Royalty in the history of his country to have studied in a school. He is the first member of the Royal family to have obtained a university degree. He is the first to have worked as the commander of his own ship, the first to have captained a sports-team, the first to pilot helicopters and supersonic jet aircraft.

His Royal Highness Prince Charles, the Prince of Wales is also colonel in chief of the 2nd King Edwards VII's Own Gurkha Rifles. HRH, the Prince of Wales is the patron of the Gurkha Welfare Trust established in 1969 for the relief of poverty among Gurkhas and their families when they leave the British Gurkha Army.

Prince Charles who also supported present 'Maiti Nepal', a rehabilitation home and orphanage for the HIV-affected women due to trafficking in Kathmandu by selling his paintings.

and

KKKG could be the first to write this book on kike or khukuri.

Meaning of life for me

Give something to this world
Each flower that blooms and dies,
Each plant that grows and dies,
Says to you and says to me
We also will die one day.
Each flower that blooms and gives
Fragrance and dies
Each plant that grows
And gives fruits and dies,
Says to you and says to me
We also have to give something to
This world before we die.

I love Nepal because I have had the chance to be born in her soil.

I Thank 'Veen' Lord!!! I have done my duty!

Jai 'Kike' !

Some clicks from my life:

WHERE ARE THEY NOW?

The photos above were taken in 1951 by the War Memorial in Sungie Patani. They show a group of Gurkha boys who had been selected as a 'Fast Stream' with the eventual aim of them attending RMA Sandhurst. The instructor in the first photo is Sgt Ronnie Coatsworth RAEC who worked at the 6 GRTW and TDBG. If you are one of these boys, or can identify any of them, then please contact the Editor.

IDENTIFICATION/WHERE ARE THEY NOW?

FIRST PHOTO (SITTING FROM LEFT):-

KULRAJ LIMBU (Retired as Maj (QGO) EX GTR (NOW settled in Kathmandu.

AKAL SING THAPA (RMA Sandhurst/Rtd as Maj) EX 6GR Now with GRH/Brunei.

KAMANSING RAI (SELF) Rtd as Maj (QGO) EX 10th PMO GR Now ILAM/Kathmandu

JOGINDRA SING GURUNG (Rtd as Capt (QGO) EX QGE NOW GRH/Kathmandu

STANDING (FROM LEFT):-

SETHMAN RAI (our head boy then) Rtd as WO2 now settled in India. (Ex QGE)

SGT RONNIE COATSWORTH RAEC (our Instructor then) Alas "GURUJI" Where R you now?

MANGAL SING GURUNG (Rtd as SGT/SSGT?) EX QGE. I met him in Pokhara re employed as plant op in 1985.

SECOND PHOTO - Standing second row second left KAMANSING RAI (myself).

21139079 Boy KamanSing Rai
Boxing Training in Sungai Patani, Kedah North MALAYA. 1953.
I am a Champion Boy Boxer of 1953. [Photo found 12 Mar 2009]
Baneswore
Kathmandu
KVKG

PHOTO Taken at Boys Training Company at Sungai Patani Kedah North Malaya in 1952, 21139079 (self) as a Champion Boys boxer of that year receiving a Champion Boy Boxer's Trophy from MGBG. Major General of the Brigade of Gurkhas.

21139079 Boy Kaman Sing Rai
Boys Training Company
Sungai Patani Kedah North-
Malaya 1952/3
Self Second Row left 1st boy.
Section Commander CPL
Kanak Bdr Ghale 2/2nd GR.
Photo found on THE MALAYAN MONTHLY, August 1959 P/8. (Year, Page)
12 Mar 2009
Kathmandu.
KRRG.

The Boy's Company of the Brigade of Gurkhas was formed in 1950 as a continuation of the old Indian Army Boy's Company. Its object: To give boys an Army education and basic instructions in drill, physical training, fieldcraft and games so that they readily absorb more technical and advanced instruction in military subjects. The boys, recruited at 14 and 15 years of age.
Captain F.W. Rawding RAEC.
[illegible] $10 per week, at 17 years [illegible]. recalled By KRRG

Admin & Tactics Course Hongkong
July 1972.

RECCE PLATOON (IN FIJI ISLAND JUNE-JULY 1971.
PLATOON COMD LT WANGDI LAMA 2I/C KKKG Then LT QGO-
RETD MAJOR (QGO) IN 1984 KIKES Reseacher-KKKG.

KKKG 1967 As a chief umpire Demonstrateg
a Reverse slope Defensive position in HongKung.

Recent photos of the Author